Vimal Katiyar

Sustainable Polymers for Food Packaging

Also of interest

Food Contamination by Packaging.
Migration of Chemicals from Food Contact Materials
Rodríguez Bernaldo de Quirós, Lestido Cardama, Sendón, García Ibarra, 2019
ISBN 978-3-11-064487-6, e-ISBN 978-3-11-064806-5

Advanced Composites -Volume 10
Biodegradable Composites.
Materials, Manufacturing and Engineering
Kumar, Davim (Eds.), 2019
ISBN 978-3-11-060203-6, e-ISBN 978-3-11-060369-9

Nano-Safety.
What We Need to Know to Protect Workers
Fazarro, Trybula, Tate, Hanks (Eds.), 2017
ISBN 978-3-11-037375-2, e-ISBN 978-3-11-038621-9

Industrial Biotechnology.
Benvenuto, 2019
ISBN 978-3-11-053639-3, e-ISBN 978-3-11-053662-1

Vimal Katiyar

Sustainable Polymers for Food Packaging

An Introduction

DE GRUYTER

Author
Dr. Vimal Katiyar
Indian Institute of Technology Guwahati
Department of Chemical Engineering
Guwahati 781 039
Assam
India
vkatiyar@iitg.ac.in

ISBN 978-3-11-064453-1
e-ISBN (PDF) 978-3-11-064803-4
e-ISBN (EPUB) 978-3-11-064463-0

Library of Congress Control Number: 2020931499

Bibliographic information published by the Deutsche Nationalbibliothek
The Deutsche Nationalbibliothek lists this publication in the Deutsche Nationalbibliografie; detailed bibliographic data are available on the Internet at http://dnb.dnb.de.

Cover image: Adrienne Bresnahan / Moment / Getty images
Typesetting: Integra Software Services Pvt. Ltd.
Printing and binding: CPI books GmbH, Leck

www.degruyter.com

Preface

The book addresses the past, present and future prospects of biobased and/or biodegradable polymers in food-packaging applications, as well as their importance and advantages over fossil polymer-based packaging materials. Further, the book also examines the current commercial overview of biobased and/or biodegradable polymers and nanocomposites, and the structure–property relationship required for various advanced applications.

Chapter 1 introduces the growing field of biobased and biodegradable polymers and their use and benefits as sustainable packaging materials, along with the classification and importance of biodegradable packaging. A glimpse of the past, present and future prospects of packaging, and advances in polymer-based packaging, including origin and advantage of biobased and biodegradable polymers for food packaging, is presented. The subsequent chapters provide a basic understanding on various biobased and biodegradable plastics used in food packaging.

The current scenario and commercial status of biobased and biodegradable polymers in food packaging are discussed with a statistical comparison of various packaging materials such as glass, metals and polymers. Comprehensive information is presented in Chapter 2 regarding sustainable polymers in relation to production statistics, including production cost and the expected growth of biodegradable polymers as food-packaging materials.

Chapter 3 discusses various biopolymers including cellulose, chitosan (CS), gums and silk, which are available in abundance in our surroundings and can be extracted from available biomass for food-packaging applications. In this section, biopolymers based on their origin, methods of extraction, processing capabilities and other specific properties in relation to food packaging have been discussed in details. This category of polymers mainly include polysaccharides, proteins and other biobased polymers. In addition, we have explained the role of these biopolymers in enhancing the properties of existing food packages.

Chapter 4 covers polymers whose precursors are obtained from renewable resources and are transformed into polymers known as bioderived polymers. These polymers can be synthesized by chemical or biochemical or both approaches. Among all the biopolymers that fall in this category, polylactic acid (PLA) has shown enormous potential to replace a wide variety of the petrochemical-based conventional polymers and is produced commercially at large scale for food packaging and other applications. Therefore, we have elaborated recent case studies regarding PLA-based bionanocomposites with special emphasis on their application in food packaging.

Chapter 5 discusses about biobased polymers that are obtained by fermentation process using microorganisms. The representative polymer in this category is polyhydroxyalkanoate (PHA), a type of biobased polymer that can be produced at commercial scale. This chapter discusses on the production, processing and their application in food packaging through case studies.

https://doi.org/10.1515/9783110648034-202

Chapter 6 gives a detailed discussion on the essential properties for food packaging materials such as thermal properties, mechanical properties, gas barrier properties, morphological and optical properties, which are the important factors for the development of sustainable plastics to be used for targeted food packaging applications. This chapter also extends discussion on the appropriate match of degradable plastics for the preservation of perishable food and the techniques towards *in situ* characterization of the shelf-life of food via active food packaging.

Chapter 7 discusses recent advancements in biodegradable polymers for food-packaging applications, in terms of technology and product development, taking into account various sustainable non-toxic natural biopolymer-based nanofillers, including those derived from starch, cellulose, CS, silk and gum. It is worth mentioning that the bioplastics filled with the above bionanofillers produce a new class of bionanocomposites with significantly improved properties that are desirable for food packaging applications. This chapter also includes the development strategies of biodegradable polymeric foams in order to develop degradable packaging that is cost competitive with fossil fuel-based packaging.

The current selling price of biodegradable plastics is deemed to be expensive. *Chapter 8* focuses on the use of potential blends of biodegradable and non-biodegradable polymers in food packaging applications, along with case studies, to provide direction to the students and young researchers to investigate the utilization of degradable polymers in combination with existing packaging.

Chapter 9 covers the important laws, regulations and legislations, and different protocols, including environmental assessment and health and safety regarding the use of biodegradable polymers for food packaging applications. Special attention has been given to the legislations regarding migration of various polymer additives inside packaged food.

Biopolymer-based edible coating for food-packaging applications is the subject of *Chapter 10*, underlying the benefits of edible coating over conventional petroleum-based plastics for food packaging applications. This chapter contains details regarding the essence of edible coatings based on polysaccharides, proteins, lipids and composite materials in combination with various active agents and their applications for the preservation and shelf-life improvement of perishable food items.

Chapter 11 discusses about the various trends on end-of-life product management including recycling, reuse of waste food packages and subsequent composting. In-depth discussion has been carried out on recycling of non-biodegradable and biodegradable packaging polymers, reuse and final composting behavior of various biodegradable polymers.

Chapter 12 presents the authors' viewpoint on the development of versatile biodegradable polymers in food packaging. This chapter highlights the shortcomings of biodegradable polymers and the possible modifications required to improve their properties in order to replace conventional plastic materials for food packaging applications, along with the scope of future research on the utilization of biodegradable polymers.

Contents

List of Contributors

The author would like to acknowledge the following colleagues for their cooperation and support in creating this book.

Chapter 1
Kiran Kumar Gali
Department of Chemical Engineering
Indian Institute of Technology Guwahati
Guwahati, Assam-781039
India

Purabi Bhagabati
Department of Chemical Engineering
Indian Institute of Technology Guwahati
Guwahati, Assam-781039
India

Vimal Katiyar
Department of Chemical Engineering
Indian Institute of Technology Guwahati
Guwahati, Assam-781039
India
vkatiyar@iitg.ac.in

Chapter 2
Purabi Bhagabati
Department of Chemical Engineering
Indian Institute of Technology Guwahati
Guwahati, Assam-781039
India

Umesh Bhardwaj
Department of Chemical Engineering
Indian Institute of Technology Guwahati
Guwahati, Assam-781039
India

Vimal Katiyar
Department of Chemical Engineering
Indian Institute of Technology Guwahati
Guwahati, Assam-781039
India
vkatiyar@iitg.ac.in

Chapter 3
Akhilesh Kumar Pal
Department of Chemical Engineering
Indian Institute of Technology Guwahati
Guwahati, Assam-781039
India

Neelima Tripathi
Department of Chemical Engineering
Indian Institute of Technology Guwahati
Guwahati, Assam-781039
India

Rahul Patwa
Department of Chemical Engineering
Indian Institute of Technology Guwahati
Guwahati, Assam-781039
India

Tabli Ghosh
Department of Chemical Engineering
Indian Institute of Technology Guwahati
Guwahati, Assam-781039
India

Prodyut Dhar
Department of Chemical Engineering
Indian Institute of Technology Guwahati
Guwahati, Assam-781039
India

Medha Mili
Department of Chemical Engineering
Indian Institute of Technology Guwahati
Guwahati, Assam-781039
India

Vimal Katiyar
Department of Chemical Engineering
Indian Institute of Technology Guwahati
Guwahati, Assam-781039
India
vkatiyar@iitg.ac.in

https://doi.org/10.1515/9783110648034-204

Chapter 4
Arvind Gupta
Department of Chemical Engineering
Indian Institute of Technology Guwahati
Guwahati, Assam-781039
India

Medha Mili
Department of Chemical Engineering
Indian Institute of Technology Guwahati
Guwahati, Assam-781039
India

Tabli Ghosh
Department of Chemical Engineering
Indian Institute of Technology Guwahati
Guwahati, Assam-781039
India

Vimal Katiyar
Department of Chemical Engineering
Indian Institute of Technology Guwahati
Guwahati, Assam-781039
India
vkatiyar@iitg.ac.in

Chapter 5
Prodyut Dhar
Department of Chemical Engineering
Indian Institute of Technology Guwahati
Guwahati, Assam-781039
India

Vimal Katiyar
Department of Chemical Engineering
Indian Institute of Technology Guwahati
Guwahati, Assam-781039
India
vkatiyar@iitg.ac.in

Chapter 6
Surendra Singh Gaur
Department of Chemical Engineering
Indian Institute of Technology Guwahati
Guwahati, Assam-781039
India

Tabli Ghosh
Department of Chemical Engineering
Indian Institute of Technology Guwahati
Guwahati, Assam-781039
India

Vimal Katiyar
Department of Chemical Engineering
Indian Institute of Technology Guwahati
Guwahati, Assam-781039
India
vkatiyar@iitg.ac.in

Chapter 7
Narendren Soundararajan
Department of Chemical Engineering
Indian Institute of Technology Guwahati
Guwahati, Assam-781039
India

Shasanka Sekhar Borkotoky
Department of Chemical Engineering
Indian Institute of Technology Guwahati
Guwahati, Assam-781039
India

Vimal Katiyar
Department of Chemical Engineering
Indian Institute of Technology Guwahati
Guwahati, Assam-781039
India
vkatiyar@iitg.ac.in

Chapter 8
Siddharth Mohan Bhasney
Department of Chemical Engineering
Indian Institute of Technology Guwahati
Guwahati, Assam-781039
India

Prodyut Dhar
Department of Chemical Engineering
Indian Institute of Technology Guwahati
Guwahati, Assam-781039
India

Vimal Katiyar
Department of Chemical Engineering
Indian Institute of Technology Guwahati
Guwahati, Assam-781039
India
vkatiyar@iitg.ac.in

Chapter 9

Narendren Soundararajan
Department of Chemical Engineering
Indian Institute of Technology Guwahati
Guwahati, Assam-781039
India

Vimal Katiyar
Department of Chemical Engineering
Indian Institute of Technology Guwahati
Guwahati, Assam-781039
India
vkatiyar@iitg.ac.in

Chapter 10

Tabli Ghosh
Department of Chemical Engineering
Indian Institute of Technology Guwahati
Guwahati, Assam-781039
India

Vimal Katiyar
Department of Chemical Engineering
Indian Institute of Technology Guwahati
Guwahati, Assam-781039
India
vkatiyar@iitg.ac.in

Chapter 11

Naba Kumar Kalita
Department of Chemical Engineering
Indian Institute of Technology Guwahati
Guwahati, Assam-781039
India

Melakuu Tesfaye
Department of Chemical Engineering
Indian Institute of Technology Guwahati
Guwahati, Assam-781039
India

Purabi Bhagabati
Department of Chemical Engineering
Indian Institute of Technology Guwahati
Guwahati, Assam-781039
India

Vimal Katiyar
Department of Chemical Engineering
Indian Institute of Technology Guwahati
Guwahati, Assam-781039
India
vkatiyar@iitg.ac.in

Chapter 12

Gourhari Chakraborty
Department of Chemical Engineering
Indian Institute of Technology Guwahati
Guwahati, Assam-781039
India

Purabi Bhagabati
Department of Chemical Engineering
Indian Institute of Technology Guwahati
Guwahati, Assam-781039
India

Vimal Katiyar
Department of Chemical Engineering
Indian Institute of Technology Guwahati
Guwahati, Assam-781039
India
vkatiyar@iitg.ac.in

Kiran Kumar Gali, Purabi Bhagabati and Vimal Katiyar

1 Sustainable polymers for food packaging: an introduction

The world's population is increasing day by day. As per the UN report, it is 7.3 billion at present and is set to grow to around 10.9 billion by 2050, where in it will become very difficult to feed the required nutrients to humans at least three times a day [1]. Therefore, storage of food with proper safety and its continuous and economical supply in adequate quantities to each person is the crucial factor where food-packaging technology plays a vital role. The ultimate purpose of packaging is to protect the food from surroundings by maintaining the quality and shelf life of the food product. The food packaging industry addresses the demands of the commercial, legal and communication barriers along with convenience tamper indication and traceability as the secondary barriers, by adopting cost-effective methods. Effective packaging materials and suitable packaging processes are the most important factors toward ensuring the quality and freshness of food at different stages of storage and distribution.

Packaging negates the chances of contamination of food items and also minimizes the wastage of foodstuff. Packaging of food items is the most important criterion, which protects it from various physical damage during transportation, preserves it from various unwanted chemical and atmospheric environment and extends the shelf life of packaged food items. Without effective packaging, it will be difficult for modern food industries to distribute bulk amount of raw and processed foods to distant areas of their marketing with a motive to supply the required nutrients uniformly to each stakeholder of the society, without any spoilage of food. In the subsequent sections, a detailed discussion will be made on different types of packaging with respect to food items and different packaging materials. Furthermore, the discussion will be extended to various plastic materials such as petroleum and biobased, differentiating biobased (in relation to their origin) and biodegradable (in relation to their end) plastics and their positive impact on the environment.

1.1 Types of food packaging

Packaged food products are mainly available in polymeric flexible film-based bags, wrappers, boxes, trays, caps, bottles, edible packaging materials and so on. Packaging helps in protecting the form, shape and texture of some delicate food

https://doi.org/10.1515/9783110648034-001

items, in addition to preventing the loss of flavors or aroma. It is a mandate that the packaging material must not affect the nutritional value and quality of a packaged food product. Along with the storage capacity, packaging in proper form is necessary to carry out safe transportation across longer distance in large quantity.

The basic functions of packaging are to:

1. contain the packaged material,
2. preserve the packaged material from outside harmful environment,
3. transport the packaged material safely from the production warehouse to the consumers,
4. provide accurate information about the packaged material to the consumers.

Based on these primary functions, packaging can be classified into three simple categories. For better understanding of the readers, Figure 1.1 illustrates the types of packaging used in the food sector in terms of its demand in the market.

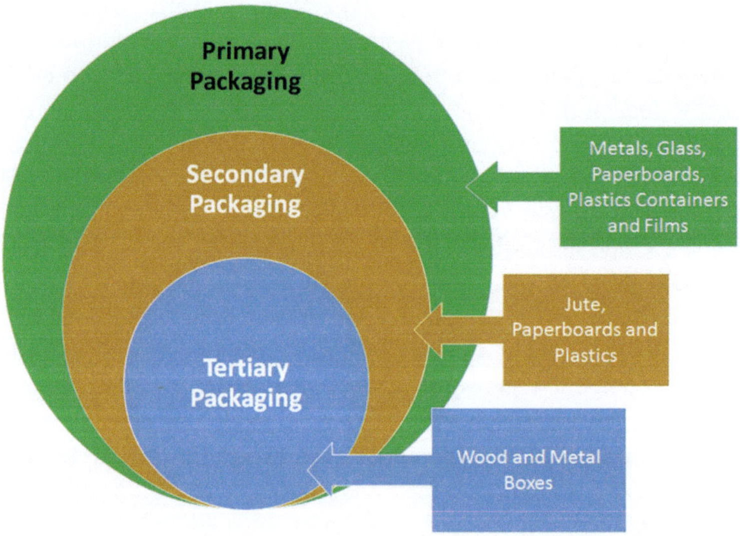

Figure 1.1: Types of food packaging in terms of its quantitative demand.

1.1.1 Primary packaging or sales packaging

This is the packaging that wraps the food items that are in direct contact with the packaging. Hence, this packaging material has much required functionalities in terms of retaining property of food items without causing any problem. Besides the action of preservation of food quality and shelf life, the packaging material should also be non-reactive with the packaged food items. Based on the type of food to be packaged, the packaging materials can be grouped into metals and its alloys, glass, polymeric containers and films and others.

1.1.2 Secondary packaging or group packaging

This is the packaging that consists of a number of primary packages regardless of whether it is sold in bulk or in loose to the consumers. Paperboard cartons and boxes made of biobased biodegradable jute bags are some of the secondary packaging materials. Also, hard and tough plastic containers, trays and boxes fall under this category of packaging.

1.1.3 Tertiary packaging or transport packaging

The functions of tertiary packaging are mainly the transportation of the primary or secondary packaged material to the customers. Another purpose of this packaging is to prevent physical damage due to inappropriate handling or transportation of grouped packages. Large wooden or metallic boxes are the most commonly used tertiary packaging materials to serve the purpose.

However, it is important to understand that different food items require different types of properties in its packaging material. Specifically, such diversity can be observed in case of primary packing materials as foods directly come in contact with the packaging materials. Hence, importance must be given to the primary packaging materials so that it does not affect the chemical, physical and aroma along with the shelf life of the packaged food items. Furthermore, detailed discussion will be made on various food-packaging materials used in the market.

1.2 Materials for food packaging

The food-packaging industry uses glass, paperboards, aluminum, tin, steel and other metals and plastics in various forms for flexible and rigid use [2].

1.2.1 Glass material

Glass is the most common and the oldest ever food-packaging material used in the form of containers, bottles, bowls, trays, cups and so on. In food packaging applications, glass is mostly used in combination with some other materials like metal, cork, plastic, rubber and others as a single unit. The type of glass enclosure depends on the nature of the food items that would be stored. For example, foodstuffs with sensitivity toward outside atmosphere or those that need sterilization by heat, will be enclosed in containers with rubbery gaskets. Foodstuffs like vinegar have a tendency to react with metals or plastics; hence, glass bottles are the perfect packaging materials due to their inertness toward such food items. Similarly, various medicines in liquid form are prescribed to be stored only in glass bottles. The inertness of glass is assumed due to its chemical structure, which consists of strong three-dimensional Si–O bonds. Also, glass being a completely amorphous solid with very small pore does not allow small gaseous molecules to pass through it and is perfect to store foodstuffs with aroma. Glass has several advantageous characteristics to be used as a potential material for the storage of various food items. It is 100% recyclable, transparent and completely see-through and is chemically inert. However, the major disadvantage in glass packaging is its highly brittle nature, which restricts its usage in several areas specifically related to transportation or rough handling. Besides being very brittle, glass pots become very heavy based on their structure and cannot be molded into critical shapes and sizes for various other types of packaging.

1.2.2 Metals

From the 1900s, metal cans have been used to store various foodstuffs across the world. The carbonated and non-carbonated drinks are the primary uses of metal cans as food packaging materials. Many a times foodstuffs are filled in the metal can in their hot state or in already cooked form to improve its shelf life without adding preservatives. While most of the food cans are packed at ambient pressure and temperature, beverages, beer and other carbonated drinks are packaged under pressure. The list of metals used for food packaging are aluminum, steel, tin, etc. Aluminium foil is being used in the household kitchen for wrapping food items due to its antimicrobial activity.

1.2.3 Paperboards

Paperboards are produced using bleached or unbleached natural fibers of cellulose. Various non-toxic chemical additives are added during the processing stage of paperboards to add certain functionalities in it. It can also be made out of recycled papers and pulp. Cartonboards made from paperboards are commonly used for packaging foodstuffs like liquid edibles, dry foods, frozen foods and fast food. Corrugated board is mostly used in applications of direct contact with food (e.g., pizza and burger boxes) and as secondary packaging materials. However, low molecular weight volatile and non-volatile additives and external contaminants can migrate through the packaging into the foodstuffs. Mostly mineral oils, phthalates, photoinitiators, etc. are the commonly used food additives that migrate through paperboard packages.

1.2.4 Polymer-based food packaging

The advantages of using polymeric materials for food packaging include the following: its easy processability and moldability; it can be formulated into any shape and size and hence offers considerable design flexibility. Polymeric food packaging materials are inexpensive and lightweight along with a wide range of physical and optical properties. The heat sealability and printability of most of the polymers is a boon to the marketing strategies of the packaged food items and can be integrated into the production line. Hence, the combination of good mechanical, thermal, processability and barrier properties along with the lightweight of the material are much effective to meet the standard requirement of packaging for customers. Basically, polymers can be classified into two categories based on its origin: (1) petroleum or fossil based and (2) biobased. More discussion on the two categories will be further made in the subsequent sections along with their advantages and lacuna in food packaging applications.

1.3 Petroleum-based plastics for food packaging

The plastics made by polymerization of monomeric units offer several advantages for food packaging applications since the plastic fluid is easily moldable and offers flexibility in design and processing into different shapes, structures and sheets. Due to the cost-effectiveness, lightweight, transparency, chemical resistance, heat stability and many more functional advantages, the usage of plastic in food packaging sectors is increasing continuously [3]. The most widely used food packaging plastic materials are polyolefins and polyesters; the other prominent materials include polystyrene (PS), polyvinyl chloride (PVC), poly(ethylene vinyl alcohol) (EVOH) and polyamide, which are among the most utilized 30 types of plastic packaging materials [4]. Polyethylene (PE) and polypropylene (PP) which come under the category of polyolefins are very attractive not only because of their barrier properties, chemical resistance, strength and stability but also due to their processability, reuse and recyclability. PE is of two forms: high density (HDPE) and low density (LDPE). The first one is stiff and strong and is used for the packaging of milk, water, juice and so on. The second one is flexible and easily sealable; hence it is used for bread, squeezable and frozen food packaging. Polycarbonate (PC), polyethylene terephthalate (PET), polycaprolactone (PCL) and polyethylene naphthalate (PEN) are the predominant polyesters obtained from ester monomers of carboxylic acid and alcohol. PET exhibits good barrier properties to gases, especially carbon dioxide and oxygen [5]. This has made PET as the best choice for packaging in many beverage industries. Polycarbonate is a clear, durable, heat-resistant material used for manufacturing water bottles and sterilizable baby bottles. It can release bisphenol when it comes in contact with harsh detergents for cleaning purposes, which is a potential health hazard [6]. PEN is a superior polymer than PET because of its high glass transition temperature. PVC is used for meat packaging because of its thermoforming property. PVC is used in flexible monolayer film packaging. PS, a brittle and hard polymer with a low melting point, is used for protective packaging applications. Polyamide, known as nylon, has similarities to PET. EVOH is an effective barrier to oil and oxygen but it is moisture sensitive and thus used in multilayered films.

1.3.1 Negative impact of petroleum-based plastics in food packaging

In today's packaging industry, use of several petroleum-based plastics and the combination of these materials to improve functional properties, consumer desires, disposal of the material and its degradation rate are posing numerous environmental, health and economic concerns. The availability of petroleum-based raw material and its cost of production, fluctuating oil prices, degradable capacity of the manufactured plastics and accumulating waste cause a serious environmental threat. One of the major issues of petroleum-based plastics is their permeability variations to light, vapors, low molecular weight molecules and gases. The need to reduce carbon emission and manage the problem of plastic litter disposal, has led to the search of new characteristic materials with antimicrobial properties. The problem of migration of particles challenges the existence of petroleum-based plastics for food packaging, as serious health concerns related to monomer residuals of plastics and other components like stabilizers and plasticizers in food cause a lot of panic to consumers. Their unpredictable impact on future generation's health aspects is also one of the issues challenging the existing food packaging industry based on petroleum-derived plastics. The food packaging demand is increasing day by day and a wide range of factors are constantly influencing this demand, which will have long-term effects in terms of environment. Therefore, food packaging and its disposal impact on the environment is studied in terms of life cycle assessment (LCA). Most of the advances in the past few decades had been dominated by plastic material mainly derived from fossil resources [7, 8].

1.3.2 Importance of biobased and biodegradable plastics for food packaging

Food is a dynamic system with limited shelf life. Packaging is done to contain, handle and protect the product from surroundings throughout the shelf life to maintain the quality and safety of food [9]. Finally, they are discarded as waste after usage. Disposal of these huge amounts of petroleum-based plastic-packaging wastes is also one of the major issues, which needs to be addressed via the techniques of degradation and composting, the diminishing of fossil resources, scarcity of raw material and increase in the cost, creating a lot of uncertainty in packaging sector. The food quality expectations, increasing awareness on health and other lifestyle aspects of consumers are posing new challenges to food packaging technology stakeholders. High cost of fossil fuels and their limited resources, waste management issues, effect on human health, environmental effects that ultimately pose long-term pollution problems and several other reasons are the prime concerns that has lead to an increased attention on the development of biobased and biodegradable food packaging. Clear environmental and sustainability benefits of biodegradable polymers are the main reasons for the surge in biobased food packaging.

1.4 Biobased plastics in food packaging

The quest to reap maximum benefits from plastics without compromising on the economic aspects, healthy environment and social acceptance, is the main concern of biobased food packaging. This section focuses on the state-of-the-art advantages and eco-friendliness of biobased food packaging in comparison to the existing plastic food packaging materials. It clearly explains us how the quality and safety of food is to be protected and disposal of the same in a systematic way by sustainable practices. The chapter also describes the various synthetic polymers that are used in food packaging and their disposal issues, contamination of food and its carbon footprints. The one more informative concept in this book is about the addition of various fillers and additives in biobased food packaging materials reach to the standards of synthetic polymer food packaging in terms of softness, lightness, transparency and mechanical properties. Clear environmental benefit in terms of fast CO_2 (carbon dioxide) fixation along with their possible biodegradation benefits, leads to the development of biobased polymers. On the other hand, fossil feedstock-based polymer development processes may lead to slow CO_2 fixation, and their complicated waste management protocols are the main reasons for the increased focus on the development of biobased food packaging materials.

1.4.1 Advantages of biobased food packaging

In order to meet the specific packaging requirements, maintain the quality and safety of food as a prime concern and also balance the environmental-friendly aspects, there is an increasing attention towards new packaging materials, which emphasizes the key factors of sustainability, that is, economic, social and environmental [10]. The growing demand for new food-packaging materials that satisfies people's requirements with reduced cost and optimized performance with ecofriendly technologies, provided the confidence for the development of many biobased materials derived from renewable feedstocks. It also explains about the biodegradation aspects for sustainability in order to cater the needs of the future generation. The LCA (life cycle analysis) is also discussed a little bit to fulfill the 3R approach, that is, reduce, recycle and reuse. The biobased food packaging are said to possess the following advantages, which suits the needs of the present generation.

- Independence from fossil fuel – can be manufactured repeatedly without depleting resources
- Entirely made from renewable sources – biobased feedstocks
- Reduce, reuse, recyclable – efficient circular economy
- Integrates the needs of present and future-sustainability
- Reduction of CO_2 emissions by 30–70% – healthy environment
- Environmental friendly and biodegradable – Green legislation
- Effective and safe throughout its life cycle – LCA

Nowadays, food-packaging sector aims to reduce the environmentally-associated problems by replacing the petroleum-based products with the biobased products from plant or plant waste, and biodegradable materials with antimicrobial properties. There are numerous applications for these materials in the food-packaging sector.

1.4.2 Classification of biobased plastic for food packaging

The biobased packaging materials are broadly classified into three sections based on their origin [11–14]. The first one is the natural polymers. These polymers are abundantly available in nature in the form of biomass; majority of them are obtained from agricultural resources. Polysaccharides, proteins and lipids are directly derived from these biomass feedstock. Starch, cellulose, lignin, pectin, alginate, chitin, agar and carrageenan are the various polysaccharides while whey, soy, casein and gluten are proteins derived from animal and plant sources. The second one is the microbial polymers. These polymers are produced by microorganisms through fermentative utilization of carbon substrates. Poly(3-hydroxybutyrate) and poly(hydroxybutyrate-*co*-hydroxyvalerate) are classified under polyhydroxyalkanoates; xanthan gum and pullulan are also microbial polysaccharides. The third one is the bioderived monomer polymers. These polymers are synthesized from bioderived monomers, which are produced by the fermentative route using carbohydrate feedstocks. Poly(lactic acid) is one of the most promising biopolymers synthesized from lactic acid bioderived monomers, which can substitute HDPE, LDPE and PET [15]. Figure 1.2 describes the classification of biobased and petroleum-based plastics in a nutshell in food packaging applications.

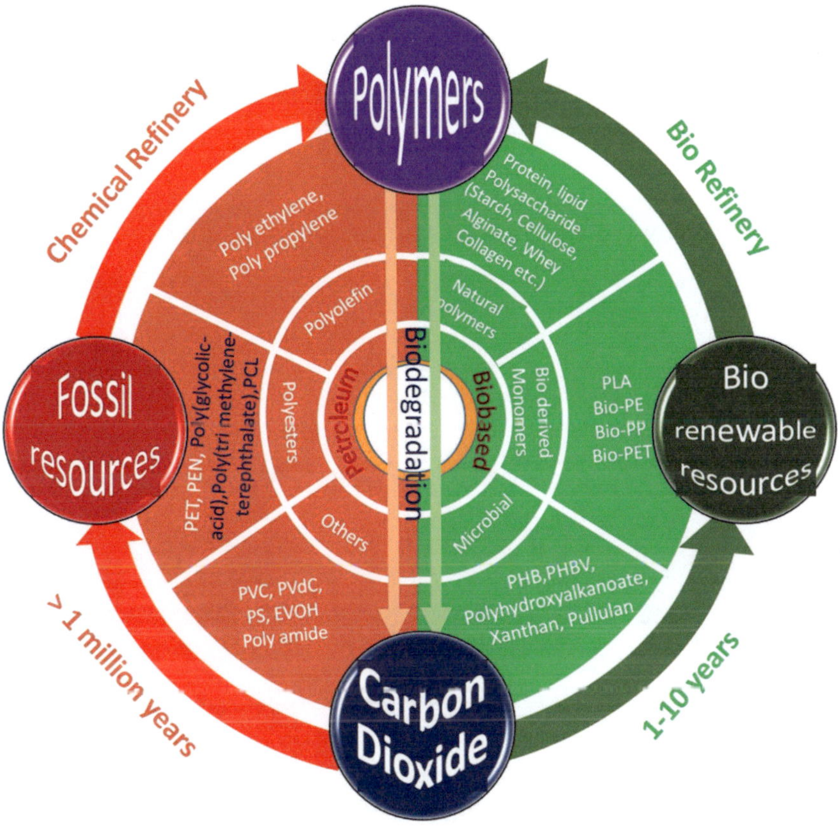

Figure 1.2: Classification of biobased and petroleum-based food packaging materials [2–18].

1.5 Differentiating biobased and biodegradable food packaging

The usage of plastics in food packaging is posing serious environmental threat, especially when these are not properly taken care of through efficient waste disposal systems such as recycling and reuse; however, eco-friendly manufacturing industries are also focusing towards development of degradable polymers, which will be suitable for sustainable food packaging. Plastic degradation is achieved either by microorganisms (known as biodegradation) or by ultraviolet light (known as photodegradation). The complete conversion of organic chemicals, that is, natural polymers into simpler compounds, thereby joining the elemental cycles by the action of microorganisms is known as microbial biodegradation. The attention given to biodegradable food packaging is a very attractive approach for reducing the plastic waste by composting them. Figure 1.3 explains about the biobased, biodegradability, LCA concepts and thus the sustainability aspect. There is a misconception that biobased and biodegradable materials are the same. But, it is not always true. Biobased signifies the origin of materials, whereas biodegradability signifies the end of its life cycle.

Figure 1.3: Food packaging life cycle concept in terms of biobased and biodegradable aspects.

Strictly speaking, the material having the capability to completely convert into carbon dioxide and water by biological degradation along with physical, thermal and chemical changes is considered as a biodegradable food packaging material. This important aspect also plays a key role in determining the LCA and ultimately the carbon footprint value, which decides the fate of the material for food packaging applications. LCA is a systematic method to evaluate the adverse impacts on the environment and human health throughout the entire life cycle of a packaging material. LCA is an important tool to quantify the characteristics and values of the product, where the data is collected and interpreted to know the economic and environmental burdens of the system at each stage of the process, that is, feedstocks, processing, finishing, marketing, distribution, usage, reuse, recycling and disposal. However, LCA is not a decision-making tool but rather gives insights for better decisions. There are various LCA reports indicating the impact of packaging and its waste on climate change, human toxicity, cancer effects, ecotoxicity and abiotic depletion, and the results are favorable toward biodegradable food packaging materials [16–18]. The following are the different assessment stages giving the scope of LCA studies:

- Stages covered under raw material to product manufacture is cradle-to-gate assessment
- Stages covered under raw material to product manufacture and thereby distribution and disposal is cradle-to-grave assessment
- Stages covered under raw material to product manufacture and thereby distribution and renewed resource (i.e., material and energy recovery) is cradle to cradle (zero impact) assessment

1.6 Conclusion

Though the applications of these biobased polymers are attractive, their usage is limited in food packaging due to their high production cost and limitations of functional properties as compared to that of competitive petro-based varieties. The renewability and less carbon dioxide emission factors have been a great driving force towards the exploration of biobased polymers as a substitute for the conventional polymers. Thus, a lot of focus is laid on the utilization of biobased polymers for developing various industrial-based products. This has led towards the research to improvise the properties of the biobased poymers by the addition of biofillers toward commercialization. In the current scenario, food packaging holds a potential market for the exploration of these biobased materials. The main reason behind the growing popularity of these polymers is that they are of low cost, can be easily processed and converted into desired shapes and also offer a broad spectrum of tailor-made properties. There will be a detailed discussion in the coming chapters about the common biobased polymers followed by a discussion on their properties, their synthetic routes of production and their potentiality as packaging materials. Later, the discussion will be further expanded in terms of the addition of biofillers and the enhancement in the properties that ultimately favors the production of biobased biodegradable food packaging materials to sustain the quality and quantity of food as well as benefit the environment in an affordable manner for the future generations.

References

[1] Department of Economic and Social Affairs, World Population Prospects: The 2012 Revision, United Nations, Population Division, New York, United States of America, 2013. (Accessed at June 28, 2017 https://population.un.org/wpp/Publications/Files/WPP2012_HIGHLIGHTS.pdf)
[2] Marsh, K., & Bugusu, B. Food packaging-roles, materials, and environmental issues. J Food Sci 2007, 72, 39–55.
[3] Lopez-Rubio, A., Almenar, E., Hernandez-Muñoz, P., Lagarón, JM., Catalá, R., & Gavara, R. Overview of active polymer-based packaging technologies for food applications. Food Rev Int 2004, 20, 357–387.
[4] Lau, OW., & Wong, SK. Contamination in food from packaging material. J Chromatogr A 2000, 882, 255–270.
[5] Willige, RV., Linssen, JP., Meinders, MB., Stege, HV., & Voragen, AG. Influence of flavour absorption on oxygen permeation through LDPE, PP, PC and PET plastics food packaging. Food Addit Contam 2002, 19, 303–313.
[6] Vom Saal, FS., & Hughes, C. An extensive new literature concerning low-dose effects of bisphenol A shows the need for a new risk assessment. Environ Health Perspect 2005, 113, 926–933.
[7] Petersen, K., Nielsen, PV., Bertelsen, G., Lawther, M., Olsen, MB., Nilsson, NH., & Mortensen, G. Potential of biobased materials for food packaging. Trends Food Sci Technol 1999, 10, 52–68.
[8] Siracusa, V., Rocculi, P., Romani, S., & Dalla Rosa, M. Biodegradable polymers for food packaging: A review. Trends Food Sci Technol 2008, 19, 634–643.
[9] Otles, S., & Otles, S. Manufacturing of biobased packaging materials for the food industry. Acta Sci Pol Technol Aliment 2004, 3, 13–17.
[10] Miller, SA. Sustainable polymers: Replacing polymers derived from fossil fuels. Polym Chem 2014, 5, 3117–3118.
[11] Stevens, ES. Green plastics: an introduction to the new science of biodegradable plastics, 41 William Street Princeton, NJ, USA, Princeton University Press, 2002.
[12] Avérous, L. Biodegradable multiphase systems based on plasticized starch: A review. J Macromol Sci Polymer Rev 2004, 44, 231–274.
[13] Bordes, P., Pollet, E., & Avérous, L. Nano-biocomposites: Biodegradable polyester/nanoclay systems. Prog Polym Sci 2009, 34, 125–155.
[14] Morris, G., & Harding, S. Polysaccharides, microbial: Encyclopedia of microbiology., Schaechter M. ed., Massachusetts, USA. Academic Press, 2009, Vol. 3, 482–494.
[15] Peelman, N., Ragaert, P., De Meulenaer, B., Adons, D., Peeters, R., Cardon, L., Van Impe, F., & Devlieghere, F. Application of bioplastics for food packaging. Trends Food Sci Technol 2013, 32, 128–141.
[16] Harbec, A. Lactic acid production from agribusiness waste starch fermentation with Lactobacillus amylophilus and its cradle-to-gate life cycle assessment as a precursor to poly-

L-lactide, Doctoral dissertation, École Polytechnique de Montréal. (Accessed at June 28, 2017 at https://publications.polymtl.ca/312/)
[17] Ferreira, S., Cabral, M., Da Cruz, NF., Simões, P., & Marques, RC. Life cycle assessment of a packaging waste recycling system in Portugal. J Waste Manag 2014, 34, 1725–1735.
[18] Degruson, ML. Biobased packaging: Reference module in Food Science, USA, Elsevier Publishers, 2016, 1–4. (Accessed at June 30, 2017 at. https://kundoc.com/pdf-biobased-packaging-.html)

Purabi Bhagabati, Umesh Bhardwaj and Vimal Katiyar

2 Biobased and biodegradable polymers for food packaging: commercial status

2.1 Introduction

Modern society is not only satisfied by the use of innovative and advance technologies but is also focused on greener approaches for minimizing the after effects of the technologies on the environment. However, packaging, being one of the major sectors affecting the environment, is being dealt with much seriousness, and research has exponentially increased on that subject. Packaging material has a share of around 45% out of the total plastic products manufacturted, and can have direct or indirect impact on the environment [1]. Therefore, a proper waste disposal management approach is required to protect the environment, wherein the conventional food packaging can be strategically replaced with the new class of sustainable plastics. These plastics will be derived from biobased feedstock and will have the capability to disintegrate after their service life, and hence will be degradable in nature. Such class of food packaging can be considered as sustainable food packaging. This chapter discusses the statistics of production and utilization of different food packaging materials annually, in-line with the market demand. The section will also include the possible packaging applications of the biobased packaging, the cost of production and the expected growth of biobased food packaging materials in the market.

2.2 Current scenario of food packaging

Global packaging industry mainly includes packaging markets for food, beverages, cosmetics and healthcare; wherein food packaging comprises almost 50% of the total packaging market. By 2020, with the continuous increase in population, food packaging requirement would increase exponentially in-line with the increasing production demand. Due to population explosion, the markets for food packaging are exponentially growing in the developing nations and excellent food packaging markets are expected to develop by 2025.

The nonbiodegradability of the conventional commodity plastics including polyethylene (PE), polypropylene (PP), polyethylene terephthalate (PET), polystyrene (PS), polycarbonate, nylon and so on, used in food packaging applications, has potentially turned out to be a major threat to the environment. Therefore, proper waste management strategies such as recycling or on-demand use of biobased, biodegradable plastics are highly encouraged. However, it is appropriate to compare the basic know-how and understanding of industrial practices toward traditional

https://doi.org/10.1515/9783110648034-002

uses of conventional plastics as it is necessary to introduce the concept of sustainable packaging. Additionally, there is a developing mandate to the corporates regarding sustainability reporting as per the government and market regulations. Such call for sustainable reporting to the industries is placing new responsibilities on the corporate sector, which inevitably will result in the inclusion of biobased or biodegradable materials in packaging of products. There is a universally accepted consensus among packaging experts on "sustainable packaging," which provides valuable contribution to the economic, environmental and social sustainability by protecting products, preventing waste, enabling efficient business conduct and providing end users with the benefits of the products without compromising with the environmental hazards. Therefore, a wide range of plastic packaging must be developed based on its compatibility with the food products, which is a critical need for the fulfilment of sustainability strategy of any company. Hence, the foremost aim of the packaging industries is the development of low-cost smart packaging along with bringing biodegradability as the key issue.

Plastics have the highest market in the packaging sector with 49% of the market share [2] as shown in Figure 2.1. Paper and board packaging have a higher share of almost 40% of the total packaging sector as compared to glass and plastic packaging, which have a total consumption share of 20% and 19%, respectively, as shown in Figure 2.2. Over the last few decades, significant advancements in science and technology have been observed in the area of plastic packaging applications, particularly in food packaging [3].

The conventional plastic packaging materials are mostly non-biodegradable, petroleum-derived precursor polymers such as PE, PET, PP and PS. These are also termed as commodity plastics that actively contribute to waste generation, thereby

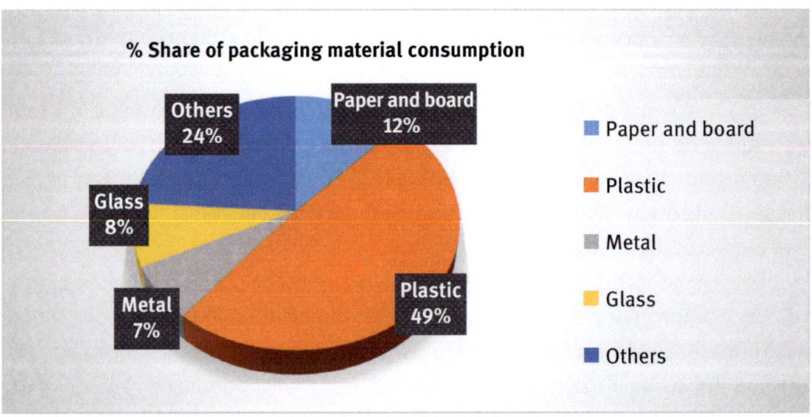

Figure 2.1: A statistical overview about the consumption of various types of packaging materials in food packaging applications.

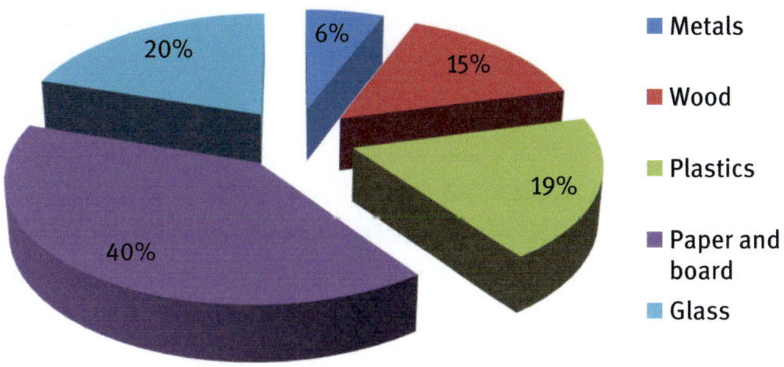

Figure 2.2: Statistical estimation of different packaging waste generated (in terms of weight percentage, wt%).

increasing pollution and deteriorating to the environmental balance of the ecosystem. Packaging statistical data of the 28 European Union (EU) member states indicate that the waste generated from plastic packaging has increased from 14.11 million tons in the year 2005 to 15.05 million tons in the year 2012, with an increase of ~6.7% over the years [4]. The demand of packaging materials shall further increase with the population explosion in the coming decades with an expected population rise of about 8.5 billion by 2030 and 9.7 billion by 2055 as compared to ~7.35 billion of the present world's population [5].

Hence, the use of non-biodegradable conventional packaging by a large population may lead to environmental issues with major contribution to municipal waste. As per the statistical data given by Organization for Economic Cooperation and Development (OECD) in 2013, an average of 530 kg of waste is generated by a single person per year in the OECD countries [6]. Total municipal waste generation would increase by 1.38 times, that is, 900 million tons per year by 2030 as compared to the year 2005 showing 653 million tons per year in the OECD countries [6, 7]. As per the statistical data on packaging waste generation by Eurostat in 2015, an average of 156.8 kg of packaging waste was generated by a single inhabitant of the EU member states in 2012, out of which almost 101.3 kg of the waste was recycled. The municipal solid waste (MSW) mainly comprises of packages made up of both degradable and nondegradable materials, food scraps and other durable items like cabinets of electronic equipments, etc. Regulatory efforts were also made to control the packaging wastes, which is the major contributor of MSW. In terms of volume, food packaging accounts for almost 50% of the total packaging wastes, which is a great value if considered in terms of quantity [4]. At present, recycling is one of the major modes to deal with the treatment of packaging wastes while energy recovery, incineration and landfilling are the other modes of treatment. There are several recovery methods available, including recycling, reusing and composting. In case of food packaging applications, polymer films with multilayered structures generally contain some other polymers that have achieved optimal barrier properties for specific applications. Separation and subsequent recycling of such multilayer films are not so easy to perform and achieve. The lack of proper collection and processing infrastructure along with the lack of consumer interest are the major reasons for the poor recovery of food packaging polymer wastes. However, the growing environmental awareness since more than a decade has caused relentless rise in the demand for biodegradable packaging polymeric materials. Thus, either proper waste management approaches are required or degradability of nonbiodegradables must be achieved to clear these packaging waste materials off our environment. Hence, initiatives to educate and overcome ignorance towards the ill effects of packaging on the environment must be made a mandate in the developing as well as developed nations.

2.3 Global initiatives toward biodegradable and biobased polymers

The legislative pressure on the ban of nonbiodegradable plastic bags along with the initiatives to tackle global warming has developed tremendous demand for biodegradable plastics. According to a report, an assumption of around 67-fold increase in the consumption of biodegradable polymers is made within the year 2050 [8]. Hence, annually the growth in the demand of biodegradable plastics will be around 15% in the coming years. In specific to food packaging and food service products, they can drive enormous requisite to use biodegradable plastics, whereas packaging foams and compostable trash bags will occupy the top position in the product list in terms of volume. Latin America is the producer of one-third of the overall biodegradable plastics produced in the world [5]. Similarly, availability of biobased feedstock also provides a number of opportunities to the Brazilian industries for the production of biobased and biodegradable polymers on a large scale. The major barriers to the use of biodegradable polymers as common food packaging materials are their high price and insufficient performance. The enacted regulations in banning plastic bags in various countries have stimulated new research and investments to develop suitable alternatives of commercially available nonbiodegradable polymers. A number of biodegradable polymers and plastics are now available that have the full potential to attain the goal. However, more scientific research is necessary to incorporate some improvements in these plastics. Use of biodegradable polymers in various fields of food packaging applications is given in Table 2.1.

The concept of "sustainable development" has entered the public debate for the first time after the World Commission on Environment and Development made a significant report in 1987, with the title *Our Common Future*. It is prominently visible by the slogan of the World Packaging Organization (WPO) "Better Quality of Life, Through Better Packaging, For More People", aiming to reduce the

Table 2.1: Various biodegradable polymers and its applications related to food products.

Area of application	Products	Biodegradable polymers	References
Packaging in food	Wrapping of various types of food items, thermoplastic food containers, nets, foams, thermoformed cups and bowls	Starch-based thermoplastics, cellulose acetates, PLA, PBS, PHB	[9, 10]
Thermoformed food service/catering	Cutlery, disposable tableware, plates, cups, straws and so on	PLA, starch-based thermoplastics	[9]
Blown film carry bags/trash bags for food items	Plastic bags for collection and composting of food waste and as supermarket carrier bags	PLA, PCL, PHB, CA	[9]
Injection-molded short shelf-life products	Transparent rigid bottles, containers, coffee machine capsules	PLA, PHB, starch-based thermoplastics	[9]

CA, cellulose acetate; PBS, polybutylene succinate; PCL, poly(ε-caprolactone); PHB, poly (3-hydroxybutyrate); PLA, polylactic acid.

spoilage and retain the nutrients for nourishing the hungry through better education on packaging. One of the main objectives of WPO includes "monitor and communicate with its members concerning packaging-related issues such as consumer safety, food preservation and environmental effect" [11]. In the packaging industry, 50% of the market is captured by food packaging. Hence, US Food and Drug Administration (FDA) is more concerned about the food packaging industry to protect the food quality distribution and aims to control the migration of the contaminants from the packaging to the food [12]. FDA expects the packaging industry to follow the Code of Federal Regulations (CFR) and follow the exemptions mentioned under CFR Part 170.39, that is the threshold of regulation for substances used in food contact articles, and work on the recyclability of the packaging materials [13]. Several initiatives have been made over the past two decades to promote "sustainable packaging" rather than waste management [14]. One of the first attempts to define sustainable packaging was made by the Sustainable Packaging Alliance in Australia [15–17]. "Sustainable packaging" is coming up as a new approach, which can become more effective through the government policy and industry self-regulation. McDonalds in America is seriously working on using biodegradable packaging materials made out of grass paper that has inherent grease resistance capacity [18, 19]. European Union directives toward improvement of waste management include the seventh Environmental Action Programme as a guidance to European environment policy until 2020. Europe's 2020 strategy towards sustainable growth along with an environmental impact includes the Resource Efficiency Roadmap as a part of the resource efficiency flagship program. To limit the production of packaging waste specifically, the EU has made an act "European Parliament and Council Directive 94/62/EC" on the 20th December 1994 on packaging and its waste. It also promotes recycling, reuse and other forms of waste recovery [20]. Amending acts on packaging and packaging waste includes Directive 2004/12/EC, Directive 2005/20/EC, Commission Directive 2013/2/EC and the latest being Directive (EU) 2015/720 for the consumption of lightweight plastic carrier bags [14]. The developments on sustainable packaging has led to the initiation of the Global Packaging Project in Paris, France, in association with the Consumer Goods Forum, an association representing over 650 global retailers, manufacturers, service providers and other stakeholders across 70 businesses in November 2008 [15]. Developing countries such as India are also not behind and are poised for the growth in the packaging industry, although the Indian packaging industry has less than 5% share of the world's packaging industry [16]. Awareness on the criticality of effects by the packaging industry is well understood by developing nations also, which is quite prominent from the initiatives like "Swach Bharat Abhiyan" by the Government of India, with one of the objectives being 100% collection/processing/disposal/reuse/recycling of MSW [21]. Food packaging industry in India follows the Food Safety and Standards (Packaging and Labeling) Regulations, 2011, made by the Food Safety and Standards Authority of India, where all the packaging used for food items must follow the rules of Bureau of Indian Standards [22].

Cellulose acetate (CA) is derived from the most abundantly available biopolymer named cellulose, which was considered as a nonbiodegradable polymer until the year of 1993 [23]. Fully biodegradable CA was commercially developed by the Eastman Chemical Company, which can be used in osmotic drug delivery, coatings, food packaging and so on [24, 25]. Poly(ε-caprolactone), polyhydroxylalkanoates and polylactic acid are biodegradable polyesters. Their hydrophobicity, good mechanical property along with ease of processing and molding make them promising candidates to be used as biodegradable polymers. Currently, PLA accounts for ~47% of the total demand of biodegradable polymers, which is followed by starch-based plastics with the value approximately as 41%. Two renowned chemical companies, Cargill Dow and Mitsui Toatsu, fabricated two different types of chemical methods for the production of PLA. At present, Cargill uses a genetically engineered yeast that converts sugar to lactic acid [26]. PLA is used as food packaging material in different forms such as packaging films, thermoformed cups and bowls and short shelf-life bottles. Cargill Dow is able to form melt spun fibers for fabrics and other applications. PBS is a biobased and biodegradable polymer that can be processed into any form by small modification in its molecular structure. Another European company named SUCCIPACK® is extensively working on several projects based on developing various packaging articles like films, trays and pouches by injection molding, extrusion film blowing and thermoforming of biodegradable PBS [27]. Li et al. have patented their invention on thermoformed articles such as food or beverage cup, lid, cutlery item, foodservice item, molded tray and food storage container, by using PBS [28]. Biobased biodegradable polymers continue to attain more and more importance in the field of packaging over any other field of applications. Due to the scarcity of land and space, several countries such as China and Germany are endorsing widespread usage of biodegradable polymers in the packaging area.

References

[1] Simon, B., Amor, MB., & Földényi, R. Life cycle impact assessment of beverage packaging systems: Focus on the collection of post-consumer bottles. J Cleaner Prod 2016, 112, 238–248.

[2] Brody, AL., & Sacharow, S. Flexible packaging of foods. Crit Rev. Food Sci Nutr 1970, 1, 71–155.

[3] Bhardwaj, U., Dhar, P., Kumar, A., & Katiyar, V. Polyhydroxyalkanoates (PHA)-cellulose based nanobiocomposites for food packaging applications. ACS Sym Ser 2014, 1162, 275–314.

[4] Marsh, K., & Bugusu, B. Food packaging – Roles, materials, and environmental issues. J Food Sci 2007, 72, R39–R55.

[5] World Population Data Sheet, Population Reference Bureau, Washington, DC, USA, 2011, (Accessed June 5, 2019, at. https://www.prb.org/2012-world-population-data-sheet-2/).

[6] Environment at a Glance 2015: OECD Indicators, OECD Publishing, Paris, France, 26th October, 2015, (Accessed June 5, 2019, at. http://dx.doi.org/10.1787/9789264235199-en).

[7] Andreoni, V., Saveyn, HG., & Ede, P. Polyethylene recycling: waste policy scenario analysis for the EU-27. J Environ Manage 2015, 158, 103–110.

[8] Schipfer, F., Kranzl, L., Leclère, D., Sylvain, L., Forsell, N., & Valin, H. Advanced biomaterials scenarios for the EU28 up to 2050 and their respective biomass demand. Biomass Bioenergy 2017, 96, 19–27.

[9] Bioplastics: Types, applications, toxicity and regulation of bioplastics used in food contact materials, 2014, (Accessed June 5, 2019, at http://www.foodpackagingforum.org/food-packaging-health/bioplastics).

[10] Bioplastics in food packaging: innovative technologies for biodegradable packaging, 2006, (Accessed June 5, 2019, at https://www.iopp.org/files/public/SanJoseLiuCompetitionFeb06.pdf).

[11] Luciano, P., & Sara, L. Food Packaging Materials, Springer International Publishing, New York, NY, USA, 2016.

[12] Bhardwaj, U., Dhar, P., Kumar, A., & Katiyar, V. Polyhydroxyalkanoates (PHA)-cellulose based nanobiocomposites for food packaging applications. Food Addit Packag 2014, 1162, 275–314.

[13] Robertson, GL. Food packaging: principles and practice, CRC press, Boca Raton, FL, USA, 2016.

[14] Tencati, A., Pogutz, S., Moda, B., Brambilla, M., & Cacia, C. Prevention policies addressing packaging and packaging waste: some emerging trends. Waste Manage 2016, 56, 35–45.

[15] Dentoni, D., Bitzer, V., & Pascucci, S. Cross-sector partnerships and the co-creation of dynamic capabilities for stakeholder orientation. J Bus Ethics 2016, 135, 35–53.

[16] Ramakrishnan, R., & Gaur, L. Information Systems Design and Intelligent Applications, Springer, India, New Delhi, India, 2016.

[17] Sorensen, JPR., Sadhu, A., Sampath, G., Sugden, S., Gupta, SD., Lapworth, DJ., & Pedley, S. Are sanitation interventions a threat to drinking water supplies in rural India? An application of tryptophan-like fluorescence. Water Resour. 2016, 88, 923.

[18] London 2012: McDonald's chooses bioplastic packaging, will recycle 75% of building, environmental leader, 2012. (Accessed June 3, at https://www.environmentalleader.com/2012/07/london-2012-mcdonalds-chooses-bioplastic-packaging-will-recycle-75-of-building/)

[19] Student designs biodegradable packaging for McDonald's, inhabitat, 2009. (Accessed 1 June 2019, at http://inhabitat.com/student-designs-biodegradable-packaging-for-mcdonalds/)
[20] DIRECTIVE, H. A. T. Council Directive 94/43/EC of 27 July 1994 establishing Annex VI to Directive 91/414/EEC concerning the placing of plant protection products on the market. Official J L 1994, 227(01/09), 0031–0055.
[21] Phalke, V. Green Marketing: Complementing Swachh Bharat Mission (SBM). Asian J Multidiscip Stud 2016, 4, 167–173.
[22] Dunford, EK., Guggilla, RK., Ratneswaran, A., Webster, JL., Maulik, PK., & Neal, BC. The adherence of packaged food products in Hyderabad, India with nutritional labelling guidelines. Asia Pac J Clin Nutr 2015, 24, 540.
[23] Klemm, D., Heublein, B., Fink, HP., & Bohn, A. Cellulose: Fascinating biopolymer and sustainable raw material. Angew Chem Int Ed 2005, 44(22), 3358.
[24] Buchanan, CM., Pearcy, BG., White, AW., & Wood, MD. The relationship between blend miscibility and biodegradation of cellulose acetate and poly (ethylene succinate) blends. J Environ Polym Degrad 1997, 5, 209.
[25] Buchanan, CM., Dorschel, D., Gardner, RM., Komarek, RJ., Matosky, A J., White, AW., & Wood, MD. The influence of degree of substitution on blend miscibility and biodegradation of cellulose acetate blends. J Environ Polym Degrad 1996, 4, 179–195.
[26] Gupta, AP., & Kumar, V. New emerging trends in synthetic biodegradable polymers–Polylactide: A critique. Eur Polym J 2007, 43, 4053.
[27] Could PBS Revolutionise Food Packaging As We Know It? (Accessed 30 May 2019, at http://cordis.europa.eu/news/rcn/122643_en.html)
[28] Li, W., Tedford, RA., Thoman, BJ., & Christie, TR. inventors, International Paper Company, assignee, 2013, US8445604.

Akhilesh Kumar Pal, Neelima Tripathi, Rahul Patwa, Tabli Ghosh, Prodyut Dhar, Medha Mili and Vimal Katiyar

3 Biobased sustainable polymers for food packaging applications

3.1 Introduction

This chapter mainly focuses on the development of bioplastics, and the associated bionanocomposites, in which the bionanofillers are extracted from biomass (primary feedstock) sources, such as polysaccharides and proteins, which in turn are obtained from agricultural or marine plants and animals. All these polymers are somewhat crystalline, having an interchain hydrogen bonding effect and are mainly hydrophilic in nature due to which problems related to melt processing and performance arises, especially when used for packaging of moist products. However, these polymers have excellent packaging charecteristics in terms of gas barrier, mechanical, thermal and other properties. The chapter mainly discusses about the various available biopolymers with their sources and properties with focused applications. The chapter highlights the research outputs and role of certain biobased polymers, which includes cellulose, chitosan, polysaccharide gums and silk. The research work carried out using the above-mentioned polymers have shown that they have enormous potential to be explored in various applications including food packaging. The main motive behind this extensive chapter is to disseminate knowledge to our readers on biobased materials and showcase their prospects in the global market.

3.2 Polysaccharides

This class of polymers mainly consists of starch, cellulose, chitosan, chitin and gums, which are mainly extracted from natural resources such as plants and microbes. The ability of the polysaccharides to form cast and blown films has compelled a lot of researchers to explore them for food-packaging applications. Over the years, a lot of research has been carried out using polysaccharide-based films and coatings as potential packaging materials to extend the shelf life, and improve the nutritional and physicochemical properties of food products. However, the main limitation associated with these types of films is that they are hydrophilic in nature and are relatively stiff in comparison to synthetic polymer-based films used for packaging. In this context, more focussed research need to be carried out to find strategies to overcome the existing shortcomings and to obtain the desired

https://doi.org/10.1515/9783110648034-003

properties required for packaging applications. A brief overview of the various polymers placed under the category of primary feedstock (category 1) is given based on their synthesis, properties and ability to use cellulose, chitosan, polysaccharide gums and silk fibroin (SF, a protein source), as industrial products for various applications especially for food packaging, exploiting their enormous potential.

3.2.1 Starch

Starch is one of the most abundantly available renewable polysaccharides, obtained mainly from cereals or tubers, and is composed of carbon, hydrogen and oxygen. Starch mainly consists of amylose and amylopectin units, where in amylose mainly contributes towards the film-forming capacity of starch. These starch-based films are found to be water soluble, nontoxic, colorless, biologically absorbable, flexible, oxygen impermeable and oil resistant. However, starch-based films do not have adequate mechanical properties; thus, various approaches such as plasticization or blending with other polymers or materials or chemical modifications have to be performed to make them suitable for packaging applications. In addition, production of starch-based polymers using extruder technique involves the conversion of both mechanical and thermal energies to thermoplastic materials. Over the past few years, these starch-based thermoplastic materials have been commercialized and they occupy a major fraction of the biobased materials' production. Starch-based bioplastics are generally developed from cereals, agar, gelatin, sorghums and others.

3.2.2 Cellulose

Cellulose is a kind of fibrous and water-insoluble linear homopolysaccharide, which is primarily composed of long chains of glucose units linked by (1 → 4) glycosidic bonds. Unlike the amylose units of starch, cellulose has β-linkages within glucose units and forms β-sheet-like structures due to the complex hydrogen bonding within its molecules. Cellulose is mainly located in the protective cell wall of plants. This particular polysaccharide cannot be digested by humans due to its dietary fiber content and needs certain microbial enzyme for its break down. Unlike starch, cellulose is crystalline in nature, which shows its amorphous nature at very high temperatures, near about 320 °C. Cellulose is used as a great source for making biopolymers, having greater polymerization in secondary cell walls as compared to primary cell walls. Moreover, cellulose has a wide application in food-packaging section where it acts as a filler material in polymers due to its various barrier properties. Again, cellulose in the form of cellulose acetate is used for bioplastic production. Another kind of organic molecule that has got wide application in the field of polymer includes bacterial cellulose, which is generally produced by the actions of certain types of bacteria on cellulose. In the production of bacterial cellulose, the principally involved bacteria are *acetobacter, azotobacter* and so on. It has different properties than cellulose molecules in terms of purity, strength and moldability. Bacterial cellulose, also known as microbial cellulose, has wide application in the field of polymer and medical fields [1]. Recently, it has been used for tissue repairing applications. The subsequent section discusses about the crystalline fraction of cellulose which upon isolation is known as cellulose nanocrystals (CNCs).

3.2.2.1 Cellulose nanocrystals

CNCs are the crystalline segments of cellulose, which are usually fabricated by removing the amorphous fractions through stringent and controlled acid hydrolysis process. These derived crystalline domains have interesting mechanical properties comparable to steel in nanodimensions with transverse elastic modulus of 18–50 GPa and anisotropic elastic modulus of 140–220 GPa [2]. However, the elastic modulus of CNCs depends upon the source of biomass used for cellulosic origin, type of acid system used for hydrolysis and analytical methods selected for mechanical property prediction [3]. The intrinsic properties of CNCs such as high aspect ratio, high hydroxyl functionality, nontoxicity and biobased origin makes it a promising bionanomaterial for wide range of applications [4]. CNCs are usually of rod shape morphology with tunable aspect ratio (length/width), which depends on the source of biomass, acid system and reaction time for hydrolysis. However, CNCs with spherical dimensions have also been reported in literature [5], which depends on the extent of acid hydrolysis process being carried out on the cellulosic pulp. CNC production is a controlled hydrolysis process, in which the protonated ions released by acid impregnates into the amorphous segments and selectively degrades it into small soluble cellobiose chains. Amorphous sections of cellulose usually have more interlayer spacing and voids [6], due to which the protonated ions diffuse and impregnate the cellulose backbones easily compared to crystalline segments. However, on increasing the contact time of the protonated ions with the cellulose chains, it might swell and diffuse into the crystalline domains, thereby degrading it into smaller units. This will lead to decrease in the yield of CNCs along with the formation of polydispersed size of CNCs. Therefore, the acid hydrolysis process should be optimized on the basis of acid concentration, ratio of cellulose pulp/acid, reaction time and temperature. The optimized CNC hydrolysis process should lead to maximum yield of CNCs along with the formation of monodispersed size of CNCs. The maximum yield of CNCs reported till date in literature [7] is ~93%, which is obtained through combined hydrolysis in the presence of hydrochloric acid under hydrothermal conditions followed by acid neutralization with ammonia. However, fabrication of CNCs from biomass and renewable resources requires additional steps of pretreatment processes to remove the impurities and extract the purified cellulose. The yield and properties of the CNCs derived from the biological origin depend on several factors such as percentage of cellulose, fraction of crystalline segments, that is, degree of crystallinity, type of cellulose polymorphs in selected biomass and also on the pretreatment method. Cellulosic source from the microorganism (MO) such as algae or bacteria and higher plants such as bamboo is highly crystalline with high content of cellulosic source [8, 9]. Therefore, CNCs fabricated from such sources are usually of high aspect ratio and have comparatively higher yields. Cellulose from plants like grass usually contains high content of hemicellulose and lignin compared to the cellulose that has relatively lower crystallinity. Therefore, CNCs fabricated from such sources are usually in nanodimensions with lower aspect ratio. Moreover, there are several waste resources such as vegetable waste, bamboo shoots, newspaper, commercial pulp, waste plant products and composts that have been used for the fabrication of CNCs at IIT Guwahati as shown in Figure 3.1, where different morphologies of CNCs are shown using field emission scanning electron microscopic (FESEM) and transmission electron microscopic (TEM) studies. Moreover, the different morphologies of CNCs obtained through different acid hydrolysis systems (H_2SO_4, HCl, H_3PO_4 and HNO_3) and polymorphs of CNCs such as CNC I, CNC II, CNC II→I obtained from waste resources are also shown in Figure 3.1. These resources have very low fractions of crystalline cellulose along with high content of impurities such as metal ions, lignin and other contaminants. In our recent study, fabrication of CNCs from compost shows the presence of mixed morphology, that is, rectangular, rod-like, spherical and cone-shaped CNCs [10]. It was possible because compost contains mixed biomass of water hyacinth, cow dung, saw dust, sand and metal ion-based impurities that interfere during the hydrolysis process, leading to the formation of mixed morphology of CNCs. Industrial residues that are produced in bulk and are difficult to dispose have also been used for CNC fabrication. Bioresidue wastes, bioethanol production plants and cassava bagasse wastes from cassava starch industries have been used for the fabrication of CNCs, thereby converting waste to value-added products [11].

In the twentieth century, among the different classes of nanomaterials, bioderived CNCs have received tremendous level of attention among researchers due to their widespread applications in healthcare, biomedical devices, packaging materials and electronic devices. CNCs have been extensively used in medical services such as targeted drug delivery and immobilization of chemotherapeutic agents for treatment of diseases such as cancer due to their bio-origin and nontoxicity. Due to their high hydroxyl functionality and ability to form a percolated network structure through intermolecular hydrogen bonding, they act as a strong reinforcing agent in the polymeric matrix. Recently, CNC films have been used as a substrate for fabrication of renewable and recyclable solar cells, which are easily disposable after their service life [12]. The fabricated CNC-based solar cells were found to be inexpensive, light weight, flexible, optically transparent and most importantly with

Figure 3.1: Cellulose nanocrystals from different biomass sources, using different acids and cellulosic polymorphs, fabricated indigenously at the Centre of Excellence for Sustainable Polymers (CoE-SuSPol) at IIT Guwahati.

high performance. Along with the optical transparency, the CNC films had very low surface roughness, which is a critical requirement for reducing reverse current leakage and higher rectification ratio in dark. The fabricated solar cells had 2.7% of power conversion efficiency, which is comparable with the solar cells fabricated on the glass substrate or petroleum-based nondegradable polymers [13]. CNCs in the form of transparent nanopaper have also found application in field effect transistor, where it acted as both substrate and gate dielectric layer, with excellent channel saturation mobility and sub-threshold gate voltage swing. This electronic characteristic of the CNC-based nanopaper makes it a potential candidate for smart displays or labels as probes, radiofrequency identification (RFID) tags and as smart active food packaging-based sensors. Therefore, these bioderived CNCs can be used as a sensor probe in active packaging as well as strong reinforcement agent for improvement in the barrier, mechanical and thermal properties of the polymer nanocomposites. In our recent study, we have found that simply tuning the degree of cellulose polymorphism during polymer processing leads to significant alteration in the structural and physical properties of nanocomposites. CNCs with cellulose II polymorphism were found to have improved mechanical and barrier properties, and hence, are more compatible for the fabrication of poly(lactic acid) (PLA) nanocomposites for food packaging applications [13]. However, studies on the migration of CNCs from packaged films to food products and its health effect have been seldom studied. Recent studies on the cytotoxicity of CNCs have shown it to be nontoxic when ingested, inhaled, or when in contact with the skin or on exposure to cells within the body [14]. Further, we carried out migration studies to ensure that the biopolymer-based polyhydroxybutyrate (PHB)/CNC films (as a model system) meet the standards as per the legislations set on migration of additives/nanofillers to the food material [15]. Detailed investigation showed that overall migration values of PHB/CNC films (at the CNC loadings range of 1–5 wt%) were within the standard limits in case of both polar and nonpolar simulants [16]. Therefore, CNC-based polymer nanocomposites have potential applications to enhance the barrier, thermal and mechanical properties as well as meet the legislative standards as per the requirement of food packaging materials.

3.2.3 Chitosan

3.2.3.1 Origin of chitosan

Chitosan is the second most abundant natural polysaccharide available on earth after cellulose. It is a copolymer of two monomers, that is, D-glucosamine and *N*-acetyl-D-glucosamine as shown in Figure 3.2, which is obtained from various renewable resources such as from marine feedstocks (e.g., crustaceans, prawn, lobsters, shrimps and crabs) in large percentages and from insects (e.g., silkworms, bees and cockroaches) in a small percentages [17–19]. Approximately 40–50% of seafood is only shells, which are the waste material for seafood lovers and contain highest percentage of chitin as compared to others. Chitin is composed of *N*-acetyl glucosamine, which is extracted from the above-mentioned renewable resources by two methods: chemical method and biological method. On the basis of yield and extraction time, chemical method is preferred over biological method. The different amounts of extracted chitin from various raw materials is shown in Table 3.1. The conversion of chitin to chitosan is performed by a single-step process called deacetylation. The structure of chitosan depends on the degree of deacetylation (DD%), which is defined as the percentage of removal of acetyl group from chitin molecule. DD% depends on various factors such as concentration of chemicals used, deacetylation time and reaction temperature. Among all the parameters, reaction temperature is the limiting factor for the conversion of chitin to chitosan. Higher DD% signifies that the

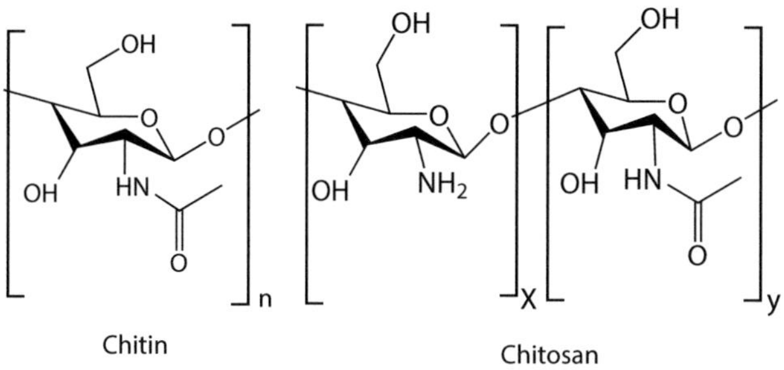

Figure 3.2: Structure of chitin and chitosan.

final extracted powder contains higher percentage of chitosan and vice versa [20]. It is noteworthy to mention that higher DD% gives lower molecular weight of chitosan due to the removal of more acetyl groups from chitin. If deacetylation time is more, conversion will be more. The main difference in chitosan and chitin is the functional group situated at the carbon–carbon interphase of the monomeric unit. The solubility and reactivity of chitosan is greater than chitin because of the presence of free amine groups in chitosan [21]. Figure 3.2 shows the structure of both chitin and chitosan.

Chitosan is approved as a food additive by the FDA because it is found to be nontoxic after oral administration to humans. In *in vitro* conditions, chitosan is degraded by the enzymes such as chitosanase, lysozyme and papain, and under *in vivo* conditions, it is degraded by lysozyme [22].

3.2.3.2 Properties of chitosan

The most important properties of chitosan are solubility, flexibility, polymer conformation and viscosity, which depends on the DD%. If the DD% is more than 50%, then it is confirmed that the prepared polymer is chitosan and it is soluble in acidic solvent. If the DD% is less than 50% or the degree of acetylation (DA%) is higher than 50%, then it is termed as chitin and it is insoluble in acidic solvents [23]. Actually, DD% and DA% are opposite to each other. The other property of chitosan is its hydrophilic nature, which is solubility in weak acidic solutions. It is not soluble in water and other chlorinated solvents. Chitosan is highly reactive than chitin because of the presence of free amine groups. It is also a good proton-conducting biopolymer electrolyte and is used in the fabrication of electrolytic cells. The versatility and applicability of chitosan in various applications are only based on the reactive amino group attached at C-2 position. It is noteworthy to mention that chitosan is insoluble at higher pH conditions because chitosan's amines are deprotonated and reactive at high pH (>6.5). On the other hand, chitosan is soluble at low pH (<6) because it is protonated at low pH and converts into NH_3^+ [20].

Table 3.1: Different sources of chitin and its yield [23–26].

Name	% Chitin
Marine sources	
Squid pens	46.23
Pink shrimp (*Penaeus duorarum*)	34.02
Indian white shrimp (*Penaeus indicus*)	30
Brown shrimp (*Penaeus aztecus*)	29.5
Mussel shell	21.32
Crab shell	16.68
Crayfish shell	15.46
Cuttlefish pens	6.12
Insect sources	
Honey bee (*Apis mellifera*)	23–32
Beetle (*Holotrichia parallela*)	15
Silkworm (*Bombyx mori*)	2–4
Silkworm (*Antheraea assamensis*)	7–8
Microbial sources	
Aspergillus niger	42
Histoplasma capsulatum	25.8–26.4
Aspergillus phoenicis	23.7
Penicillium chrysogenum	19.5–42
Trichoderma viridis	12–22
Blastomyces dermatitidis	13
Paracoccidioides brasiliensis	11
Mucor roxii	9.4
Neurospora crassa	8–11.9
Tremella mesenterica	3.7
Saccharomyces guttulatus	2.3

3.2.3.3 Limitations of chitosan

It has higher oxygen barrier properties depending on the operating conditions but its poor mechanical and weak water resistance in the presence of water and humidity limits its applications because it is hydrophilic in nature. These properties can be improved by adding plasticizers and salt, cross-linking of polysaccharides, use of suitable solvent, chemical modification of hydroxyl groups, change in pH, addition of different polysaccharides and blending with other polymers. The other drawback is its solubility because it gets dissolved only in acidic medium. It has a film-forming ability but the prepared chitosan film is brittle in nature so it can't be used in packaging applications without using any plasticizer or a sufficient modifier. One limitation also comes into picture that the vegetarian people may avoid such packaging due to its origin [20, 24, 25].

3.2.3.4 Methods of chitosan extraction

Chitosan can be extracted from renewable resources by two methods: chemical method and biological method. The fabrication or extraction of chitosan from marine and insect wastes is the most important step, which governs its various attributes such as molecular weight, purity, DD% and polydispersity index. The utilization of chitosan in different applications can be tuned with the help of these properties. The extraction methods should follow the criteria for commercialization and must be simple, time saving, energy saving, economical and eco-friendly [20, 26, 27]. As per the literature survey, the marine creatures such as shrimp, crab, shellfish and krill have 40–50% of waste biomass like heads and exoskeletons, which are composed mainly of chitin and proteins. This waste biomass can be utilized as a substrate for chitin production.

Chemical methods of extraction: The uncontrolled dumping of seafood wastes converts it into a pollutant, which may disturb the ecological balance of the environment. However, dumping of waste biomass can be controlled with the help of the seafood industries by utilizing it for the extraction of chitosan due to its versatile chemical composition. The seafood wastes are composed of protein–chitin matrix, which is broadly calcified in the presence of mineral salts to convert it into recalcitrant shells. Some strident treatments are required to extract chitin from the complicated arrangement of chitin with other available constituents. According to the conventional chemical method, a known amount of exoskeleton or shells is washed many times and chopped into smaller sizes to remove the dirt, dry flash residues, lipids and other removable impurities. Further, the dried sample is grinded into the powdered form for the ease of treatment. The demineralization step can generally be performed by acid treatment using acids such as HCl, H_2SO_4, HCOOH, CH_3COOH and HNO_3 [26]. Among all these acids, HCl is highly preferred for the removal of minerals such as catechols, Ca, Mg, K and other water-soluble impurities from the raw material as shown in Figure 3.3. But, such a strong acid may harm the molecular weight of chitin, which results in the reduction of physicochemical and intrinsic properties of chitin. The demineralization step with strong acid can completely remove the organic salts. The next step is deproteinization in which the dried demineralized powder is treated with sodium hydroxide to eliminate proteins and other organic impurities [23]. The deproteinized powder is further treated with bleaching agents such as acetone, potassium permanganate, hydrogen peroxide and sodium hypochlorite at room temperature to remove color and pigments. The obtained dried powder is known as chitin. The chitin is converted to chitosan by deacetylation process in which the acetyl groups are removed from chitin molecule. Generally, the deacetylation reaction can be performed at different time intervals at various temperatures, which can affect DD% and quality of chitosan. The deacetylation can be performed in the presence of concentrated alkali such as sodium hydroxide and potassium hydroxide, which can convert chitin into chitosan partially. Amine groups are released during alkali treatment, which imparts the characteristic of cation. The deacetylation reaction at room temperature and elevated temperature is called as homogeneous deacetylation and heterogeneous deacetylation, respectively, depending on the required nature of chitosan. Generally, the extraction of chitosan by chemical method may produce higher yield. However, it consumes higher energy and causes pollution in the environment. The chemical treatment may alleviate the waste disposal problem and the valuable protein may also get damaged, which cannot be used further for animal feed.

Biological methods of extraction: Biological method imparts more attention due to the above-mentioned disadvantages of the chemical method. The commercially available enzymes can be utilized for chitin extraction at laboratory scale. However, the production of chitin from biological method may be uneconomical and disadvantageous due to longer incubation time, lower efficiency and costly sterilization procedures. Hence, the living MOs such as *Pseudomonas aeruginosa, Enterococcus faecium, Bacillus subtilis* and *Candida parapsilosis* can be used as the economical and

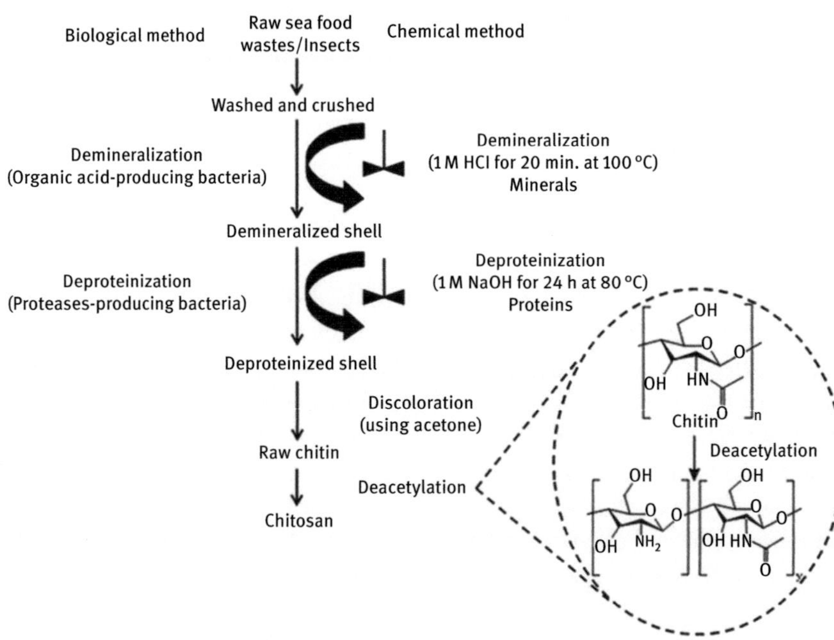

Figure 3.3: Extraction procedure of chitin and chitosan from natural resources.

attractive alternatives for chitin purification efficiently. Proteolytic enzymes are released by the panoply of MOs, which have higher stability toward organic solvents, pH and temperature. The optimization of process parameters may be one of the ways to reduce the production cost of chitin. The recovery of valuable by-products such as free amino acids, proteins, minerals and carotenoids during extraction can also help to reduce the production cost. Among the biological methods, fermentation is one of the most safe, ecofriendly and economical methods [28]. The successful development of fermentation process has triggered the industries to replace the existing traditional chemical methods. The solid-state fermentation technique can be used at industrial scale in a controlled manner to obtain desired products. It can also regenerate the natural microbiological methods, for example, ensiling and composting. Ensilation of seafood waste renders various advantages and is appropriate for the production of value-added products such as proteins, pigments and chitin. Protein- and mineral-rich liquor fraction can be converted to chitin-rich solid fraction with the help of MO-mediated deproteinization and demineralization. The biowaste deproteinization primarily occurs using proteolytic enzymes, which are produced by *Lactobacillus*. As a result, pure chitin fraction can be obtained from by-products, that is, liquor with higher percentage of amino acids and soluble peptides. The usage of *Lactobacillus* bacteria or protease-producing bacteria can replace the acid treatment step and concentrated alkali treatment step in chemical methods. Other bacteria such as *P. aeruginosa* K-187 and *B. subtilis* can also be used for the deproteinization of shrimp waste [29]. Further, biological demineralization can also be performed using various reactions such as enzyme-catalyzed reactions, microbial processes and natural probiotic reactions, which involve alcalase, *L. pentosus* 4032 and milk curd bacteria, respectively. Chitin and chitosan production using biological methods are mainly dependent on various fermentation parameters such as temperature, initial pH value, inoculum size, carbon concentration, sugar level, incubation time, carbon-to-nitrogen ratio and their optimization [28].

3.2.3.5 Functionalization of chitosan

Chitosan is a natural polysaccharide, having various characteristics such as biodegradation, immunological, antimicrobial, antibacterial, antifungal and wound healing activity in human body. Chitosan is the only polymer, which is used as a support material in tissue engineering, gene delivery, cell culture and many more. Hence, the functionalization of chitin and chitosan with various synthetic and biodegradable polymers has received great attention to overcome its limitations and increase its utilization to enhance various physicochemical and biological properties. The functionalization of chitin and chitosan can be performed using physical and chemical methods. However, chemical modification is more stable and productive than the physical modification [30].

3.2.3.6 Chemical functionalization of chitosan

The presence of NH_2 group at the C2 position in the chitosan molecule makes it an interesting polysaccharide, which is highly reactive than the OH group. The unique NH_2 group can be modified as per the required properties with improved biological functions [31]. Solubility is an important parameter to analyze the effect of functionalization. Hence, to improve the solubility parameter of chitosan in various organic and inorganic solvents, the hydroxyl groups in chitosan can be functionalized also. Basically, chitosan can be modified in a similar fashion to that of cellulose by various reactions such as esterification, etherification, crosslinking and grafting. A variety of chitosan-based products can be industrialized with unique properties such as antifungal, antibacterial, antiulcer, antiviral, nontoxic, antiacid, nonallergenic, biodegradability and biocompatibility by amino-specific chemical reactions, that is, quaternization, acetylation, alkylation, reactions with ketones and aldehydes, chelation of metals [31, 32]. Other reactions like grafting, hydrogen bonding and o-acetylation may be performed on the hydroxyl functional groups. Wang and coworkers [33] have synthesized chitosan-graft-polycaprolactone (PCL) with the help of green reaction media, that is, 1-ethyl-3-methylimidazolium acetate using ring-opening polymerization in the presence of stannous octoate ($Sn(Oct)_2$) as a catalyst under controlled reaction conditions and feed ratios as shown in Figure 3.4(a). The grafting content of PCL on both –OH and NH_2 groups of chitosan reached up to 630%. The synthesized copolymer was highly advantageous due to the properties of both the reactant polymers and could be utilized in drug delivery, tissue engineering and gene delivery [33].

Zhou and coworkers [34] have synthesized self-aggregated nanoparticles (~200 nm) of PLA-grafted-chitosan (PLA-g-CH) copolymer via protection-graft-deprotection reaction for its potential use in ocular delivery of amphotericin B (AmB) as displayed in Figure 3.4(b). The problems related to therapeutic gaps have been overthrown with the help of AmB-loaded nanoparticles based on PLA-g-CH (AmB/PLA-g-CH). Such polymer system is able to enhance sustained drug release, drug contact time and drug penetration without ocular irritation [34].

Yao and co-workers [35] have developed cytocompatible graft copolymer using chitosan and L-lactic acid with the help of grafting onto amino groups as shown in Figure 3.4(c). The crystalline structure of chitosan is disturbed after grafting due to the randomly arranged side chains after the substitution of $-NH_2$ group. The cell growth rate is faster on grafted copolymer than that of chitosan as confirmed by the static cultivation of in vitro fibroblast, which is continuously decreased with an increase in feed ratio. Hence, chitosan-grafted-lactic acid can be a promising candidate in tissue engineering [35].

Chitosan is not only used in biomedical applications but also in food packaging due to its film-forming ability. Pal and Katiyar [36] have synthesized lactic acid oligomer-grafted-chitosan (OLLA-*g*-CH) by in situ polycondensation polymerization in the absence of any catalyst as shown in Figure 3.4(d). The prepared nanoamphiphilic grafted copolymer has been successfully used as a nanofiller in PLA matrix for the improvement of oxygen barrier properties in food packaging applications. The synthesized copolymer is like a surfactant in which hydrophilic chitosan acts as micelles head and oligo L-lactic acid (OLLA) acts as micelles tail in the dimension of ~2–4 nm. More than 18 °C reduction in the glass transition temperature (T_g) of the fabricated bionanocomposite films by solution casting method confirmed the improvement in plasticization effect of PLA matrix. The oxygen transmission rate decreased up to approximately 10-folds with an increase in filler loading, which is critically required to improve the food shelf life in packaging applications [36].

Figure 3.5 shows the utilization of hydrophilic chitosan for the synthesis of hydrophobic OLLA-*g*-CH, which is blended with PLA using solution casting method and found a uniform dispersion of OLLA-*g*-CH molecules in the form of three-dimensional spherical particles in the PLA matrix, which was confirmed by FESEM and TEM analysis. The filler works as a nucleating agent and increases the nucleation density of crystals due to the presence of nanonucleating sites. The crystallization kinetics of bionanocomposite films shows the reduction in crystallization time with the increase in OLLA-*g*-CH content. The spherulitic growth of such bionanocomposites increases with the increase in filler loadings.

In another study, Peesan and coworkers [37] have developed hexanoyl chitosan (H-chitosan) (as shown in Figure 3.4(e)) by reacting chitosan powder with hexanoyl chloride in a solvent mixture of chloroform and anhydrous pyridine. The functionalized chitosan has been further utilized to draw composite fibers of PLA/hexanoyl chitosan using electrospun technique with the help of various solvents such as chloroform, dichloromethane and tetrahydrofuran. The prepared composite fibers have been compared with PLA and hexanoyl chitosan in various aspects. The morphology of spun PLA fibers has been observed to be circular in their cross section with rough pore-like surface. This pore-like surface is the result of high volatility of the solvents. On increasing the concentration of spun PLA fibers, the viscosity of the solution as well as diameter of formed fibers increases. The morphology of spun H-chitosan fibers has been observed to be flat with smooth ribbon-like surface, which is also independent from the used solvents. The multiple melting phenomena of spun PLA and H-chitosan/PLA fibers have been observed by the differential scanning calorimetry thermograms. The synthesized polymer composite fibers are highly desirable for wound dressing and tissue scaffolding [37].

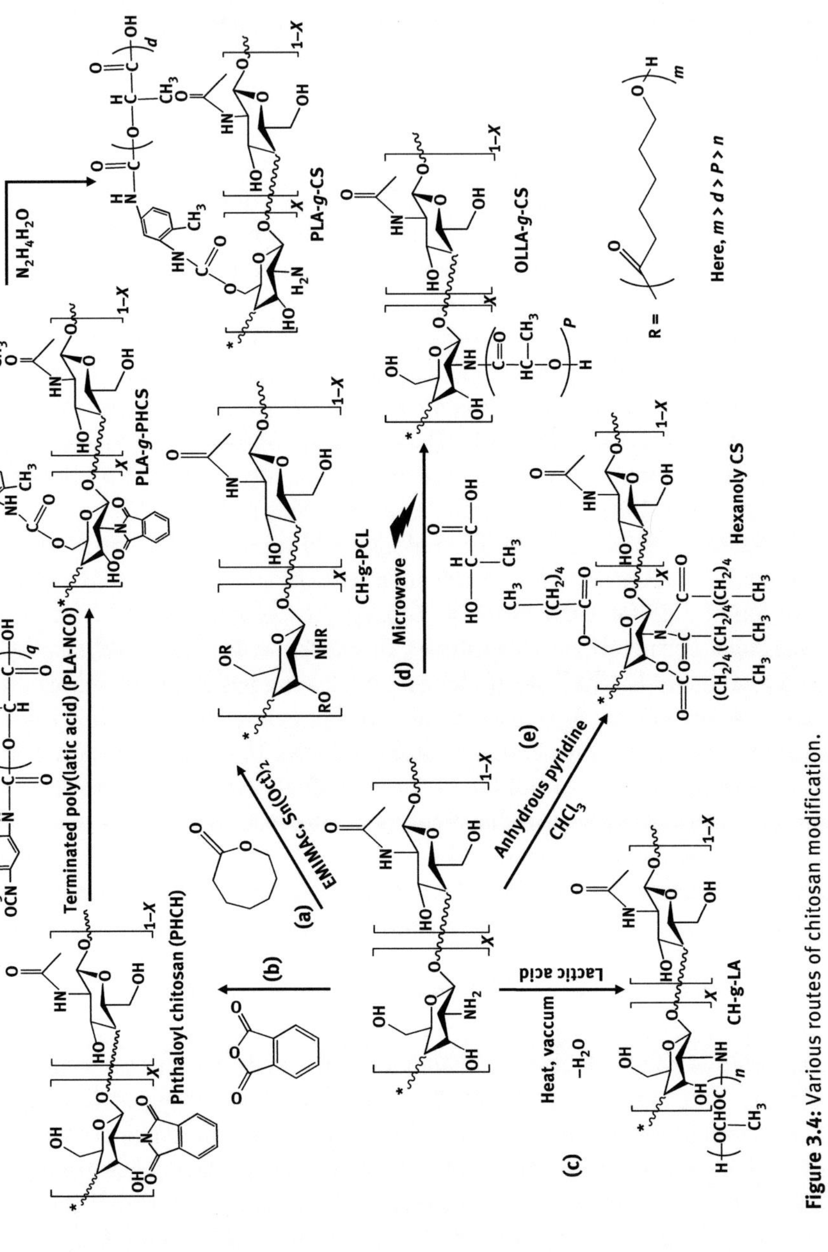

Figure 3.4: Various routes of chitosan modification.

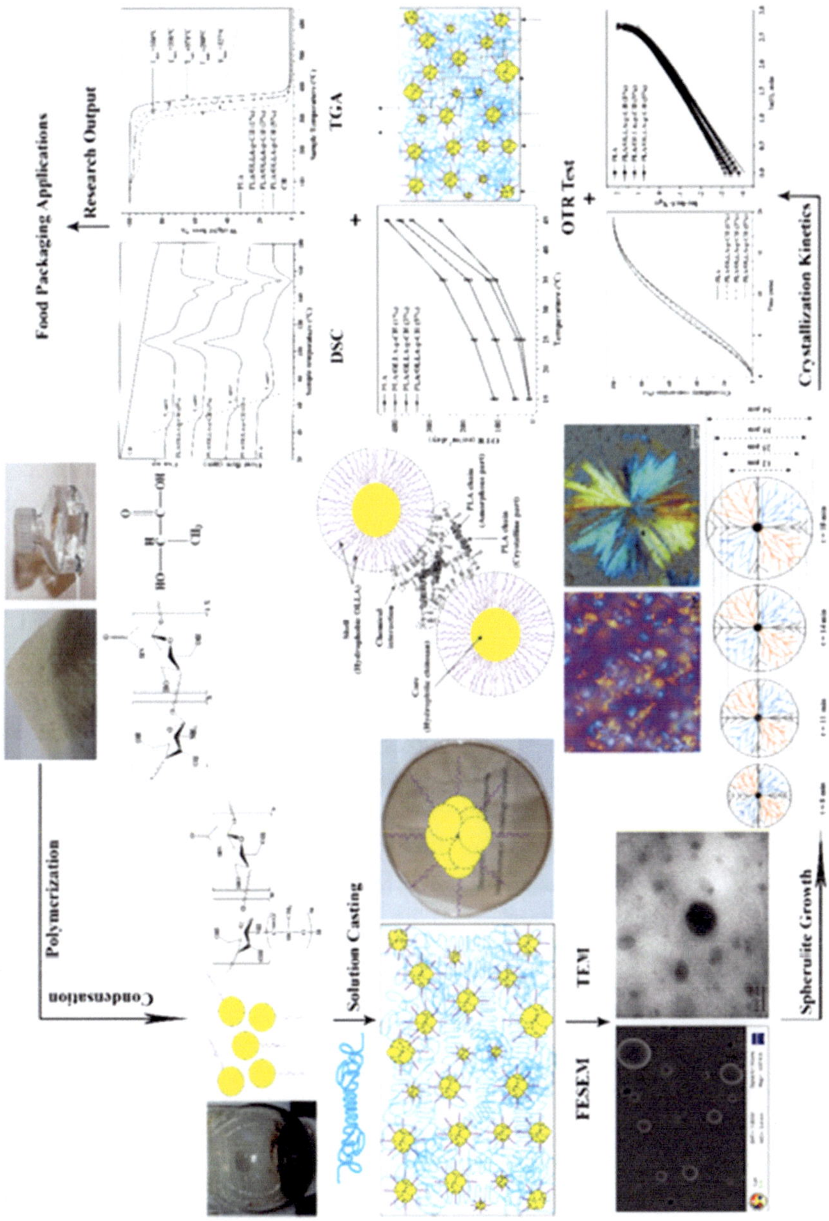

Figure 3.5: Usage of chitosan in packaging applications.

Chitosan is highly demanded for its biological properties such as antibacterial, antimicrobial and antioxidant, which have excellent applications in relation to food packaging. Such properties were tuned by Schreiber and coworkers [38]. They had prepared neat chitosan as well as functionalized chitosan films with the addition of antioxidants. The functionalized chitosan has been synthesized using gallic acid, 1-ethyl-3-(3-dimethylaminopropyl) carbodiimide and *N*-hydroxysuccinimide in different proportions during reaction, and was further used as a filler to prepare chitosan/gallic acid-grafted-chitosan films in order to verify the required biological properties for food packaging applications. According to the results, grafted chitosan has higher DPPH (α, α-diphenyl-β-picrylhydrazyl) scavenging ability (90%) and reducing power (0.51) than that of nongrafted chitosan. Gallic acid has been grafted on chitosan at C_2, C_3 and C_6 positions. C_2 position refers to the amide bond with amino group. C_3 and C_6 positions refer to the formation of ester bond. Grafting efficiency depends on the DD%, so based on the results, chitin has much lower level of total phenolic content and antioxidant activity than that of chitosan. No significant change has been observed for chitosan with 100% DDA and 80% DDA. It means that only a certain extent of amino groups and available hydroxyl groups had grafted with gallic acid [38]. Chitosan is very much compatible with other biodegradable polymers. Soares and coworkers [39] have utilized this property of chitosan and cross-linked it with glutaraldehyde and used it as a coating solvent on thermoplastic starch (TPS) and TPS/PLA sheets, which have been prepared by a thermopresser machine. The coating has been done by two different methods, that is, spraying and immersion. The TPS/PLA sheets have shown good processability during molding, good handle-ability and homogeneous appearance without cracks. A continuous covering of chitosan has been observed on the surface of coated or uncoated TPS/PLA films irrespective of the coating method. That continuous sheet prepared from immersion is more irregular than that of spray coating. The water solubility of coated TPS/PLA films with cross-linked chitosan has been observed to be less than that of uncoated sheets. In the case of barrier properties, the coated TPS/PLA films with cross-linked chitosan have higher barrier properties produced by the spraying method. Water vapor permeability of spray-coated TPS/PLA films has also been reduced to 35% than that of uncoated films. The tensile strength and elastic modulus of coated TPS/PLA films have increased than that of control films, whereas elongation at break decreased as the covalently bonded macromolecules have formed a three-dimensional network, reducing the mobility of polymer chains. Such prepared films can be helpful in packaging applications [39]. Hence, it is proved that the chitosan is highly flexible in terms of its use in various forms. It is noteworthy to mention that chitosan can also be utilized with synthetic polymers to balance their nondegradability, which can help to reduce the polymer waste and maintain the ecological balance. Vasile and coworkers [40] have developed low-density polyethylene (LDPE)/chitosan composite films using various chemicals including LDPE, chitosan or chitosan sodium montmorillonite (MMT) clay nanocomposites (CHnano) in the presence of natural antioxidant, that is, vitamin

E (VE) as well as with and without a synthetic antioxidant, that is, Irganox 1076 by melt processing technique. The antimicrobial properties have been observed after the addition of chitosan into the LDPE matrix due to the antimicrobial property of chitosan. Improved oxidation induction period has been observed by addition of VE in the LDPE matrix. The antimicrobial and thermal properties have been improved with the addition of both chitosan nanocomposites and VE into LDPE matrix due to the increase in surface charge and synergistic effect. The obtained new materials had good inhibition activity against different bacteria such as *L. monocytogenes, E. coli* and *Salmonella enteritidis.* Such kind of films can be used for food packaging applications. Human health is not affected by the incorporation of above additives into the natural products [40]. Except the film-forming ability, chitosan has another positive point, which is related to the gas barrier property in food packaging applications. Chitosan films can be considered as impermeable to oxygen at specific conditions. It can prevent oxygen to enter into the plastic packaging film. Many different types of additives are added into the biopolymer films to improve food quality, minimize microbial growth and extend shelf life. The additives may be antioxidants, antimicrobials, colors, antifungal agents and nutrients. According to Abdollahi and coworkers [41], properties such as physical, mechanical, antimicrobial and antioxidant are improved by incorporating MMT nanoclay and rosemary essential oil (REO) into the chitosan film. These properties are changed by changing the MMT weight % and REO levels. Other biomolecule-based active polymers can also be used for packaging but chitosan is more recommended because of its antimicrobial behavior and bivalent mineral-chelating ability. But an existing drawback of chitosan is that it shows poor mechanical and gas barrier properties, weak water resistance in the presence of water and humidity. These properties can be improved by adding plasticizers and salt, cross-linking of polysaccharides, use of suitable solvent, chemical modification of hydroxyl groups, change in pH, addition of different polysaccharides and blending with other polymers. Polymer clay nanocomposite (PCN) is an option that is used as conventional filled polymers because of its nanoscale dispersion ability. Mechanical and physical properties are higher in PCN than microscale polymer composites. MMT is the most useful hydrated alumina silicate-layered clay having edge-shared octahedral sheet of aluminum hydroxide between two silica tetrahedral layers. The negative charges on the surface are counterbalanced by interchangeable cations, which are typically Na^+ and Ca^{2+}.

Generally, most of the industries are taking interest in natural antioxidants and antimicrobials instead of the synthetic ones. As discussed earlier, essential oils (EO) are used for this purpose because EOs have greater activity than active compounds due to the synergistic effect. Rosemary (*Rosmarinus officinalis* L.) is the EO extract from plants that belong to Labiatae family. It has phenolic compounds such as rosmarinic acid, rosmadiphenol, rosmanol, epirosmanol, rosmadial, carnosol and carnosoic acid, which show antioxidant property; similarly, α-pinene (2–25% of composition), bornyl acetate (0–17%), camphor (2–14%) and 1, 8-cineole (3–89%) show antimicrobial properties. Sample characterization has been done by x-ray diffraction (XRD), Fourier transform infrared spectroscopy, film thickness measurement, water resistance test, water vapor permeability calculation, tensile strength calculation, surface color measurement, swelling analysis, antibacterial activity of EO and total phenol assay. From the experiments, it was observed that at low concentration (1–5%), the inorganic and hydrophilic MMT can disperse into chitosan solution very well and results in a favorable interaction with hydrophilic chitosan. Hydrogen bonding is the factor, which affects the formation of exfoliated and intercalated hybrids. It was also observed that there may be a particular arrangement in the films because of interaction of MMT and REO ingredients. Finally, it was concluded that it is a suitable example of active bionanocomposite for food packaging application [41].

3.2.4 Polysaccharide gums

3.2.4.1 Gum arabic

Gum arabic (GA) is a neutral or slightly acidic salt of a complex polysaccharide containing Ca, K and Mg cations [42]. Moreover, it is a kind of arabinogalactone protein, which is structurally similar to it [43, 44] and the main chain is a highly branched macromolecule consisting of galactose units linked with β-1,3-glycoside bonds (36–42%) (Figure 3.6(a)). The side chains present in different fractions, which differ from product to product, are formed by L-arabinose (24–29%), L-rhamnose (12–14%), D-galactose and D-glucuronic acid (16–17%). GA with maltodextrin and gelatin can be used as microencapsulating agents [45, 46]. In the principal chain, they branch off from the carbon 6 of galactose units. Molecular weight observed for GA was M_n >250,000 and M_w ~880,000 [47].

3.2.4.2 Guar gum

The term guar is derived from the word "guaran," which is a naturally occurring polysaccharide [48, 49]. Guar gum (GG) is a polygalactomannan obtained from the seeds of a Leguminosae plant or the Indian cluster bean, *Cyamopsis tetragonolobus* [50–52]. GG is mainly the ground endosperm of guar beans [49]. India produces 80% of the world's GG [51].

Chemically, GG is a heteropolysaccharide composed of sugars such as galactose and mannose as shown in Figure 3.6(b) [51]. The backbone is a linear chain of (1–4)-linked β-D-mannopyranose, and short side branches are composed of (1–6)-linked α-D-galactopyranose, which is connected at every alternate mannose units [49–51, 53]. Because of the hydroxyl groups present in the molecule, they are capable of absorbing water, and their hydrophilic properties increases as a result [54]. These nine OH groups in a single unit help in forming a hydrogen bond to mineral surfaces [55].

GG is a promising component for biobased biodegradable plastics due to its low cost, non-toxicity, easy availability and biodegradability [54–57]. It is nonionic and hydrophilic in nature, and can entrap a large amount of water in the range of subzero to 78 °C [58, 59]. It has good solubility in cold water and its aqueous solution is highly viscous. It can be used in food, cosmetic and paper industry due to its capability of having high shearing rate [60]. GG and its derivatives can also be used in textile industries. Some of its limitations are batch-to-batch variation, prone to viscosity reduction upon storage, uncontrolled rates of hydration and microbial contamination [48].

3.2.4.3 Xanthan gum

Xanthan gum (XG) is an exopolysaccharide produced by the bacterium *Xanthomonas campestris*. XG is an anionic heteropolysaccharide with a primary structure consisting of repeated pentasaccharide units. These pentasaccharides are formed by two glucose units, two mannose units and one glucuronic acid unit, in the molar ratio 2.8:2.0:2.0 as shown in Figure 3.6(c) [61]. The primary structure of XG contains a cellulosic backbone (β-D-glucose residues) and a trisaccharide side chain of β-D-mannose–β-D-glucuronic acid–α-D-mannose attached with alternate glucose residues of the main chain. The nonterminal D-mannose unit in the side chain contains an acetyl functional group. The anionic character of this polysaccharide is due to the presence of both glucuronic acid and pyruvic acid groups in the side chain [62].

The average molecular weight of XG is between 2,000,000–7,500,000 and 15,000,000–50,000,000 [63, 64]. XG is produced by different bacteria listed in Table 3.2. They are grown and inoculated in a bioreactor to form XG.

3.2.4.4 Research on polymers with natural gums

Natural gums are also classified under the category of biopolymer; however, its combination with synthetic polymers as composite or as an adhesive has not been explored significantly. Abdel-Halim and coworkers prepared graft copolymer of GG and polyacrylamide for the preparation of silver nanoparticles [49]. From the UV–vis spectra, high improvement in the absorption intensity and no decrement in the peak intensity were observed. This meant that the reduction power got higher and the grafted copolymer had better stabilization efficiency. In another research paper, biodegradable hydrophilic synthetic polymer, that is,

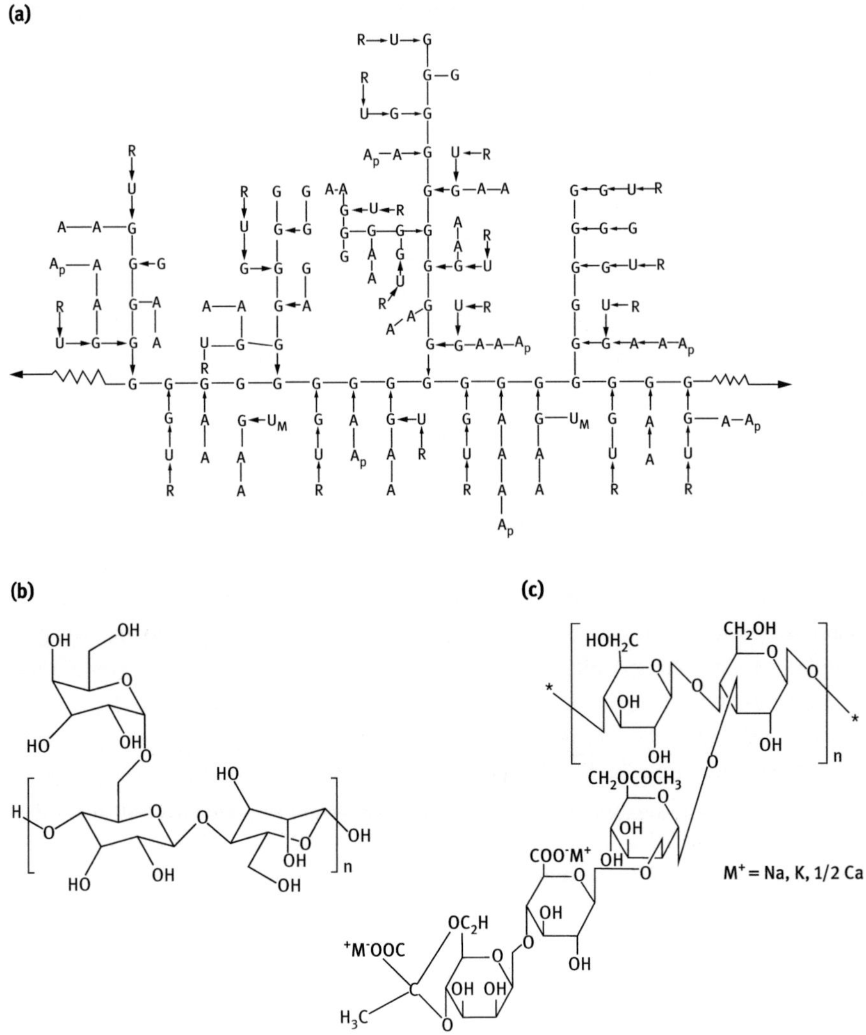

Figure 3.6: (a) Structure of gum arabic: R, rhamnose; Um, 4-*O*-methyglucuronic acid; U, glucuronic acid; Ap, arabinopyranose; A, arabinose; and G, galactose, (b) Molecular structure of guar gum and (c) Molecular structure of xanthan gum.

polyethylene oxide (PEO) was used with XG, and the blend miscibility was tested based on viscosity, ultrasonic velocity and refractive index methods [65]. Further, the addition of GA in polyvinyl alcohol (PVA) decreases the T_g along with an increase in mechanical properties [66]. PLA/PCL copolymer-biodegradable hot melt adhesive was also prepared for packaging applications. The degradation of this copolymer was found to be significant than pristine polymers [67]. Detailed comparisons are listed in Table 3.2.

3.3 Protein-based biopolymers

Proteins are one of the subclasses of biopolymers that consist of polymerized amino acids, where blocks of amino acids are linked through peptide bonds. Proteins mainly consist of a single or several peptide chains. However, the proteins can also be linked to nonpeptide components (derived protein). Structural-based protein mainly includes keratin, collagen and elastin, whereas functional-based proteins include enzymes and hormone. Protein-based biopolymers are mainly formed through condensation reaction, ring-opening polymerization and metal catalyst. Nowadays, whey protein, casein, egg, blood meal, soybean, gluten and pea protein are the commonly used protein sources for the production of biopolymer [71]. A study of combined usage of peoprotein (cheap source and low-allergic content) and glycerol (acts as plasticizer) blends decreases the intramolecular forces between the polymer chains resulting in the decrease in glass transition temperature [72]. Both vegetable and animal proteins are widely used to develop protein-based biopolymers. The quality of biopolymer formed mainly depends on the structure of protein (primary, secondary and tertiary), its hydrophobicity, cross-linking type, types of bonds, linking molecules, types of additives, plasticizers and so on. Protein-based biopolymers mainly consist of several unique types of proteins or may be blended with other biobased materials. Incorporation of other biobased materials increases the property of the biopolymer. Plasticizers help to reduce the processing temperature during extrusion process.

The most widely used techniques for the development of protein-based polymers include casting, thermo-molding, extrusion and injection molding. The processing of bioplastics using extrusion technique mainly occurs through the denaturation steps of protein. Denaturation of proteins mainly occur by thermal and chemical methods. Globular proteins need unfolding and realignment during processing. Use of plasticizers during extrusion helps to reduce the degradation rate of protein during processing. Moreover, better protein-based bioplastic can be developed by understanding the nature of proteins and its various properties. The important parameter of extrusion processing is processing temperature, solubility, chain mobility and others. Various properties of the developed biopolymer can be measured by using XRD and infrared spectroscopy. In addition, glass transition temperature can be measured by differential scanning calorimeter and dynamic mechanical thermal ability. Furthermore, the developed environment-friendly packaging materials include the use of biobased nanofillers [73], nanocomposites [74], biodegradable plastics [75] and bioplastic [76], having better properties than that of conventional plastic packaging materials. Besides this, nowadays protein-based biopolymers include the development of silkworm-based polymer, spider-based polymer, chicken feather and so on.

Table 3.2: Gum–polymer interaction.

Gum	Polymer	Process	Outcome	Reference
Guar gum (GG)	Polyacrylamide	***Preparation of polyacrylamide/guar gum graft copolymer*** GG + water in dilute sulfuric acid. Acrylamide + water + GG. Solution temperature of medium was raised under continuous stirring conditions. At the end potassium bromate and thiourea dioxide were added in the reactor and polymerization reaction was allowed to proceed. Polyacrylamide/GG graft copolymer was separated from the homopolymer via precipitation.	Better stabilization intensity was achieved in case of graft copolymers.	[49]
Xanthan gum (XG)	Gelatin (polysaccharide)	Gelatin and XG systems were prepared. Gelatin was dissolved in deionized water to prepare 10% solution at 75 °C. XG was dissolved in 4% solution at 60 °C and subsequently, sucrose was added. Then GS (glucose syrup), gelatin and XG solution were added into the sucrose solution.	The addition of XG to gelatin networks resulted in high levels of cosolutes changes in the T_g and kinetics of glass transition and glassy states.	[68]
XG	Polyethylene oxide: a biodegradable hydrophilic synthetic polymer	A dilute solution of 0.2% homopolymer was prepared and was blended in different compositions in NaCl solution. Viscosity, ultrasonic velocity and refractive index were determined at 30 and 40 °C. The required temperature was maintained in a thermostat bath with a thermal stability of ±0.5 °C.	Based on viscosity, ultrasonic velocity and refractive index methods, the polymer blend of XG/PEO was found to be miscible only when the polysaccharide XG content was 40% and above and there was no effect of temperature on the miscibility of the blend system.	[65]

XG and carboxymethyl cellulose (CMC)	Maize masa	Maize masa was mixed with CMC and XG at 5% (w/w) solid ratio and then freeze-dried to obtain dried masa samples. For preparation of masas, maize was normalized to obtain the samples that were then mixed with the hydrocolloids (CMC and XG).	Maize masa was found to be well mixed with XG and CC after treatment.	[69]
XG	Cassava starch	Cassava starch (3–5% w/w) was blended with water and the additives: acetylated or deacetylated XG (0–1% w/w), sucrose (0–2% w/w), propylene glycol (0–1% w/w), sodium phosphate (0–0.2% w/w) and soybean oil (0–0.06% w/w). Medium pH was adjusted with 50% citric acid solution or 5% sodium hydroxide solution, then heated to 75 °C with constant stirring and placed under vacuum (30 min), to remove bubbles that could become pinholes after film drying. The films were prepared in Petri dishes. Samples were stored (23 °C, 75% RH) for at least 4 days prior to testing. Since starch films have a hydrophilic character, 75% relative humidity was chosen for conditioning the experimental films in order to evaluate the material performance on high moisture content environment.	In Comparison to PVC films, lower tensile strength resistance values were observed after addition of sucrose. Increased water activity (hydrophilicity) was observed for films in the presence of sucrose, increasing the material biodegradation.	[70]
Gum arabic (GA)	Polyvinyl alcohol (PVA)	Solution casting technique as above.	The presence of PVA polymer protects the GA in the blend from degradation. The optimum ratio of GA in the blend was found to be 30% (w/w). Lower T_g of the blend can be achieved with increasing GA content. Moreover, there is an improvement in mechanical properties in terms of stress–strain behavior.	[66]

3.4 Protein biopolymer (silk)

3.4.1 Classification of silk-based polymers

The term silk refers to the wide range of continuous filaments spun by several species of Lepidoptera and Arthropoda [77]. They are used to perform varying functions in cocoons that protect larvae/eggs to spider draglines to capture nets that can trap prey [78]. Both types of silks do not have any genetic heritage in common, and their amino acid sequence compositions are different too. However, the silkworm and spider employ a similar spinning process to produce silk. Blocks of poly-Gly-Ala occur in fibroin, while blocks of poly-Ala with few Gly residues occur in spidroin [79]. The noncrystalline regions or the amorphous domains are the randomly coiled sections that are rich in glycine forming a 3_1-helix mostly parallel to the fiber [80]. The ratios of crystalline to noncrystalline regions vary between spider and silkworm silk. β Sheet content is considered to be less in spider silk (35%) when compared with silkworm silk (40–50%) [81].

Silk fiber from the silkworm comprises of two components: SF and silk sericin (SS). A native silk fiber contains two SF strands with triangular cross section, thus making silk shimmer. These are separated by a layer of SS as shown in Figure 3.7 [82]. In contrast, sericin made up of amorphous amino acids is hydrophilic in nature amounting to 25% weight of cocoon [83]. It is soluble in warm water and is removed from silk by a process called degumming, rendering it smooth texture while maintaining its tensile strength. In addition, it has several extraordinary properties aimed at protecting the silk, cocoon and pupa inside, such as oxidation resistance, antibacterial and regulating the moisture content [84].

3.4.1.1 Macro- and fine structure

The high crystallinity imparts the mechanical function, while the extensive hydrogen bonding makes the silk insoluble in almost all solvents, including dilute acids, alkali as well as water [85]. The hydrophobicity of the silk prevents water to enter the structure and imparts high packing density [86]. To strengthen the knowledge about the structure of silk, various techniques like NMR, XRD and TEM have been extensively used [87]. Fibroin is traditionally described as rigid with inextensible β-sheet crystallites embedded in a rubbery matrix as shown in Figure 3.8. X-ray diffraction measurements reveal the presence of secondary, β-sheet structures (crystal parts) resulting from the repeating motif GAGAGS [87].

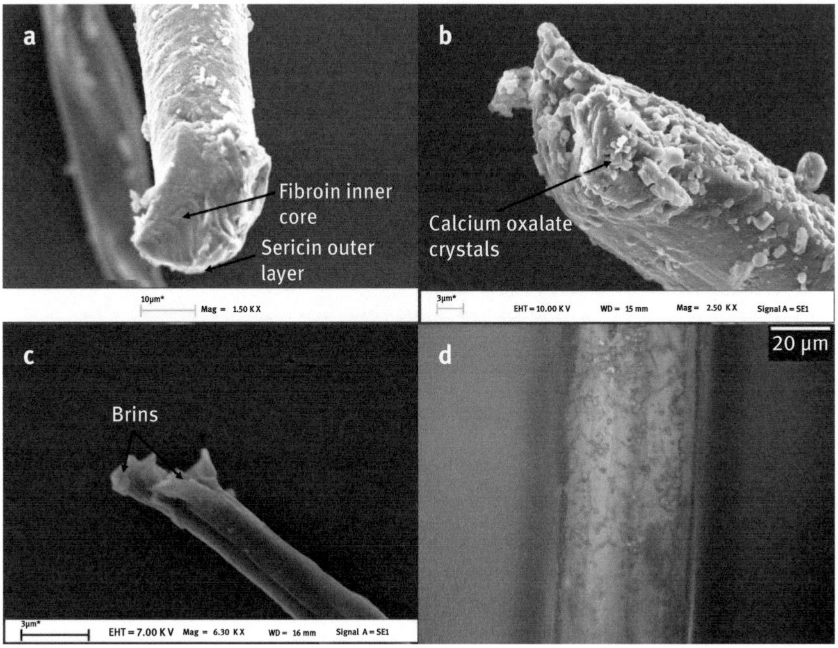

Figure 3.7: (a) Microscopic view of silk fiber with sericin outer layer; (b) microscopic view of raw silk fiber with calcium oxalate crystals over its surface, (c) microscopic view of degummed silk fiber showing two brins and (d) optical microscopic image of raw silk fiber showing high shimmer under light due to its high refractive index.

3.4.2 Processing of silk

Silk requires an annual production of 4,00,000 tons of cocoons and is used in textiles and other applications. For using of SF in packaging application, it can be processed by aqueous and solvent formulations into different material morphologies (as shown in Figure 3.9) [86].

3.4.2.1 Silk fibers

Composite fibers made from spidroin and PLA were found to have a core sheath morphology. A higher concentration of spidroin (~15%) improves the mechanical properties [82]. Composite fibers of fibroin and PVA had improves the breaking strength but were found to be inflexible [88]. Composite fibers composed of 8.5 wt% fibroin and PAN had fibroin sheath which was developed for application in textiles.

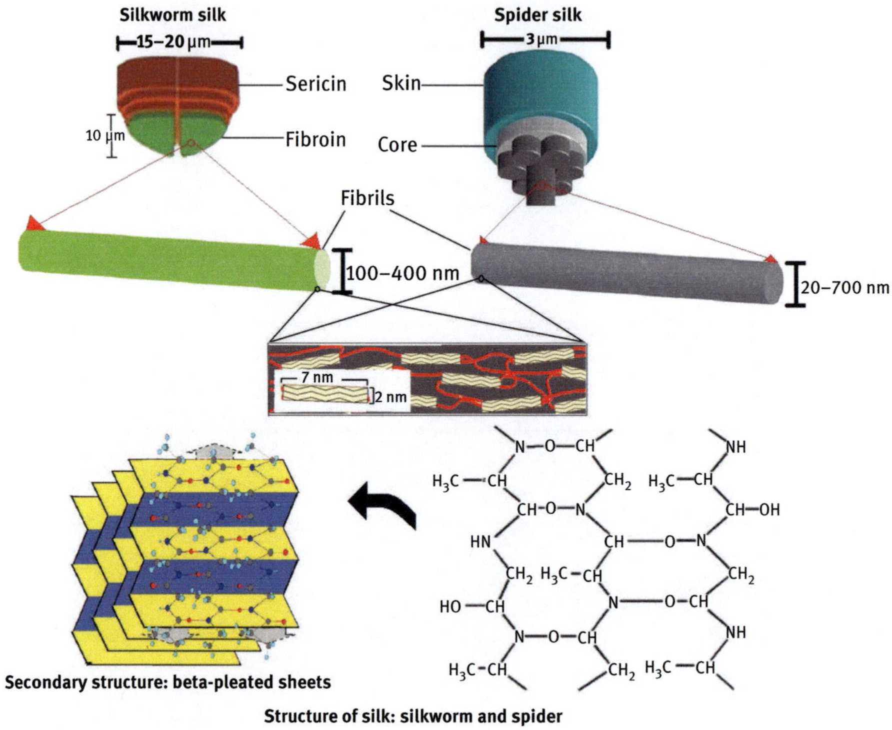

Figure 3.8: Arrangement of crystalline and amorphous domains in silk fiber.

Currently available textile engineering techniques are now being used to generate ligament replacements like the human anterior cruciate ligament [89].

3.4.2.2 Electrospun/nonwoven SF mats

Nonwoven mats are prepared by partial solubilization of native silk with formic acid or treatment with calcium. These mats have high surface area and have a rough surface enabling cell attachment [86]. Electrospun fibers can be produced in a wide range of diameters, ranging from a few nanometers to a few microns depending on the mode of processing. The hMSCs (Human mesenchymal stem cells) cultured on these mats showed attachment and spreading with lateral modulus comparable to native silk fibers (~13 GPa) [90]. Electrospun meshes with random coil structure were converted to β-sheet structures with methanol treatment, which reduces porosity due to reduction in water content [91]. Nonwoven silk mats had fibers of 30–120 nm diameter when formic acid was used as solvent. SF is blended with PEO

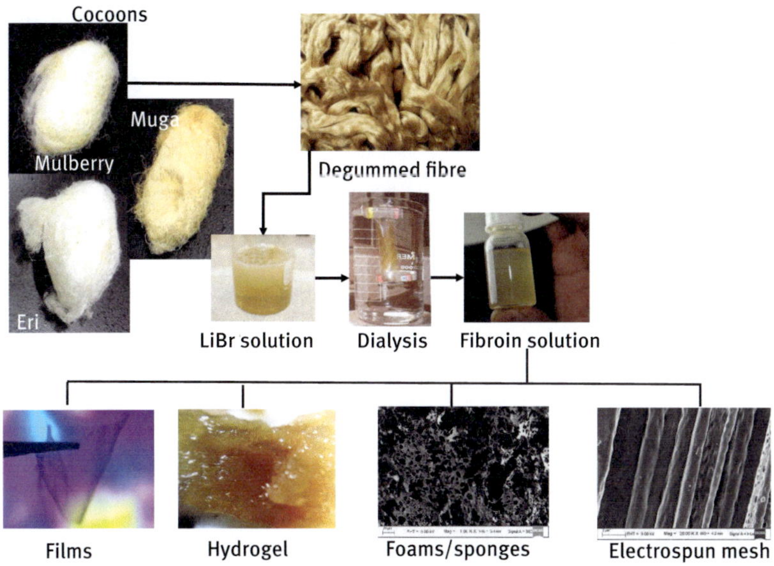

Figure 3.9: Various processing pathways available for silk.

and electrospun to produce nanofibers for delivery of morphogens like bone morphogenetic protein-2 [92]. These nonwoven mats can also be coated with polymers that are conductive such as polypyrrole and polyaniline, which imparts them electrical conductivity. This characteristic can be used to stimulate cellular growth upon application of low electric fields [93]. In addition, they may also act as smart packaging materials.

3.4.2.3 SF films

Composite films of fibroin particles homogeneously dispersed in PCL resulted in opaque films, where in opacity increased as the fibroin content was increased [94]. Composite films of fibroin and PLA were cast using dioxane as solvent and were found to interact via hydrogen bonding. The films were found to have increased tensile strength and elongation when they contained 10 wt% PLA [95]. Composite films of fibroin and PCL upon exposure to electron beam generated radicals. Subsequent studies on such systems showed that exposure to electron beams generates radicals that were responsible for cross-linking and led to enhanced flexural and tensile strength. Composite films of fibroin and poly(allylamine) were found to be more stable in water when compared with neat poly(allylamine) films that are readily soluble in water [96]. Perfect crystalline films composed of blends of *B. mori* fibroin and nylon 66 were cast from formic acid solution [97]. SF films have spin-coating, dry-cast or layer-by-layer assembly processes from organic or aqueous solvent systems when other polymers are blended along with silk [93]. Dry-cast silk films are brittle when dried with low breaking strain of ~0.5–3%, which limits its use in most of the practical applications [98]. On the other hand, ultrathin films that can be made by layer-by-layer and spin-coating techniques have the desired toughness of ~328 kJ m^{-3} and UTS of ~100 MPa, which is comparable to most conventional polymer composites [99]. By changing the silk structure, water vapor and oxygen permeability can be fine-tuned. Silk structure can be altered by treating with alcohol, temperature or radiation [100]. This structure change produces silk with variable mechanical properties and degradability. Another alternative to methanol treatment can be water annealing but it ruptures the films with low β-sheet content, which makes the films stronger [85]. Blending silk with cellulose produces transparent films with increased mechanical strength compared to silk alone [101]. Whey protein isolates blended with SS also rendered films that had increased mechanical properties and low water vapor permeation, and were both edible and biodegradable [102].

3.4.2.4 SF hydrogels

Hydrogels made from mixture of fibroin and PVA were prepared in a three-step method of cold incubation, lyophilization and rehydration. Elongation properties improved when PVA concentration was increased but the strength remained unaltered [103]. It was observed that fibroin could be effectively cross-linked with PEO, which improved both mechanical properties and water swellability [104]. Sol–gel technique produces fibroin hydrogels with interconnected network of fibroin molecules. Various techniques for inducing gelation are sonication, change of pH, change of temperature and using ionic species such as Ca^{2+} as cross-linker [105]. For hydrogels to be mechanically strong they should have small pore size with high packing density. Silk–gelatin hydrogel system showed structure transition upon temperature change, which resulted in varying mechanical and rheological properties developed for drug delivery applications [106].

3.4.2.5 SF porous sponges/foams

Composite foams are produced by freeze–drying mixtures of PLA and fibroin dissolved in dioxane. It was found that the two materials interacted via hydrogen bonding, and fibroin increased hydrophilicity of PLA foams [107]. Composite foams were prepared by freeze–drying poly(ε-caprolactone) foams filled with aqueous solutions of fibroin [108]. Moreover, the strength and elongation at break increased as the PCL content increased. Composite films of fibroin and PEO showed improved elongation while tensile strength was affected [92]. Another technique for producing ordered array fibroin sponges that are highly stable and possess β-sheet structure with required hydrophobicity was by casting aqueous fibroin solutions over polystyrene microspheres and subsequent ethanol treatment for conformation change and then treatment with toluene to dissolve the polystyrene structures [109]. Silk-based porous sponges prepared using porogens, gas foaming and lyophilization are obvious materials for tissue engineering applications as they allow for flow of nutrients and waste which are much needed for cell attachment, migration and proliferation. Fibroin sponges with their high surface area can have a porosity as high as ~97% [110]. Recent study showed promising results in healing of rat epidermis, where sponges were made with a blend of PVA, chitosan and silk [82].

3.4.2.6 Surface modification

New functionalities such as antibacterial activity, sensing functions and therapeutic effect can be introduced in SF upon introduction of functional components (drugs, nanoparticles and antibodies) [93]. It is ensured that introduction of these new functionalities does not compromise on strength and toughness. Silk surface can be functionalized by the use of available amino acid side chains (only 3.3% carboxyl groups available). Surface modification includes physical adsorption or chemical immobilization of a protein or ligand [96]. Glucose oxidase was immobilized on SF films for use as a glucose sensor [111].

3.4.2.7 Degradation

US pharmacopeia considers SF fibers nonbiodegradable as it retains more than 50% of their mechanical properties even after 2 months in vivo. The delayed biodegradability can be useful in developing smart and active packaging [86]. Silk-like other natural polymers can be degraded with the help of proteases, especially α-chymotrypsin, which can help bring down molecular weight to half in ~17 days [112]. Silk degradation can be controlled by modification in morphology, structure, mechanical and biological conditions.

3.4.3 Applications of silk

SF apart from being used in textiles is also used in biomedical field (tissue engineering, drug delivery and diagnostics). All the above-mentioned applications use its intrinsic properties like smooth texture, luster, smooth texture, easy processability and thermal stability. Most of the applications are possible because of its low cost, abundant supply and superior mechanical properties. In the following sections, we will discuss the applications of SF from the mechanical perspectives. Keeping in mind the above-mentioned advantages, one can think of using silk toward packaging usage as well.

3.4.3.1 Tissue engineering

SF possesses the required mechanical strength for tissue scaffolds, which also acts as a stable template for regeneration of tissues such as cartilage, bone and skin [113]. Silk can be processed into different morphologies with varying mechanical properties that can match both soft and hard tissues [89, 105].

Silk is an excellent suture due to its low bacterial adherence, good handling characteristics, high strength and it can be absorbed within 60 days in vivo [89]. Nonbraided silk varieties have smoother surfaces that do not provide grooves for inflammation causing substances like immune cells to adhere [114]. This same product can be used for tying/stitching of loose products replacing nylon threading and reducing the carbon footprint to some extent.

3.4.3.2 Therapeutic agent delivery

Silk can be used as a carrier for therapeutic agents such as drug/growth factors. Excellent stabilizing effects and mid-processing options along with excellent mechanical strength make SF a suitable option. This can play a major role in packaging industry for smart packaging as controlled release of essential nutrients. Cellular delivery means delivering therapeutic agents, in general growth factors encapsulated within a semipermeable membrane are diffused out to the targeted tissue [115]. SF with its high strength maintains encapsulation even when mechanical stresses are induced upon injection to the target site. The desired strength can be achieved even by methanol/water annealing, ruling out high temperature and UV treatment, which may damage therapeutic agents during processing [116]. This induces sufficient β-sheet structure required for slow release even for 10 months [117].

3.4.3.3 Optics and sensing

Fibroin has a refractive index higher than water, air and other materials, making it a suitable candidate for use as a supporting or substrate material in sensors and optics [118]. Various inherent properties such as its lightweight and biocompatibility further advocates its use in optics. Silk can be loaded with probes or dyes and it can also be used to generate patterns as optical and sensory parts in diagnostic, therapeutic, smart packaging and bioinstrumentation applications. Fibroin has been imbued with dyes/probes or patterned to generate various optical and sensing components useful in bioinstrumentation, smart packaging pharmaceutics, medical diagnostics and therapeutic applications. It is noteworthy to mention that fibroin is now used to develop optical fibers, pH, oxygen, glucose, alcohol sensors, diffraction gratings, lens arrays and so on [93].

3.4.3.4 Mechanical immobilization

Fibroin can achieve mechanical stability with aqueous processing, often without the need for chemical cross-linking or UV-curing [119]. As such, optical or sensing probes, especially biological probes (e.g., hemoglobin, myoglobin, enzyme peroxidase and antibody immunoglobulin G), can easily be incorporated into fibroin via mechanical entrapment [120]. Many smart packaging devices like RFID-like silk sensors are developed, which can now tell the current status of fruit ripening. Similarly, fibroin-based packaging sensors are available for dairy industry, which can detect milk/milk product spoilage by direct contact or by immersion. It is noteworthy to mention that all these current applications are based on edible fibroin devices [120].

3.5 Conclusions and outlook

Use of biobased polymers in the field of food packaging is increasing day by day. In this chapter, discussion on the outgrowths of research and development towards increasing the use of primary feedstock-based bioplastic has been done. The use of starch, cellulose, CNC and SF as nanofillers provides better property as food packaging. Moreover, these biopolymers with its varied morphologies and different properties hold numerous applications in the areas of drug delivery, sensors, textiles and so on. All these applications are possible due to the materials' thermal, mechanical and gas barrier properties, which make them a good natural substitute for many synthetic polymers.

References

[1] Fu, L., Zhang, J., & Yang, G. Present status and applications of bacterial cellulose-based materials for skin tissue repair. Carbohydr Polym 2013, 92, 1432–1442.
[2] Lahiji, RR., Xu, X., Reifenberger, R., Raman, A., Rudie, A., & Moon, RJ. Atomic force microscopy characterization of cellulose nanocrystals. Langmuir 2010, 26, 4480–4488.

[3] Lahiji, RR., Reifenberger, R., Raman, A., Rudie, AW., & Moon, RJ. Characterization of cellulose nanocrystal surfaces by SPM, NSTI nanotechnology conference and trade show, nanotechnology 2008, 1–3, 704–707.
[4] Dhar, P., Bhardwaj, U., Kumar, A., & Katiyar, V. Polyhydroxyalkanoates (PHA)-Cellulose Based Nanobiocomposites for Food Packaging Applications, In: Food Additives and Packaging, Eds., Vanee Komolprasert, Petra Turowski, ACS Sym Ser, 2014, Vol. 1162, 197–239.
[5] Baek, C., Hanif, Z., Cho, SW., Kim, DI., & Um, SH. Shape control of cellulose nanocrystals via compositional acid hydrolysis. J Biomed Nanotechnol 2013, 9, 1293–1298.
[6] Cowling, EB. Comparative biochemistry of the decay of sweetgum sapwood by white-rot and brown-rot fungi, No. 1258, US Department of Agriculture, 1961.
[7] Yu, H., Qin, Z., Liang, B., Liu, N., Zhou, Z., & Chen, L. Facile extraction of thermally stable cellulose nanocrystals with a high yield of 93% through hydrochloric acid hydrolysis under hydrothermal conditions. J Mater Chem A 2013, 1, 3938–3944.
[8] Elazzouzi-Hafraoui, S., Nishiyama, Y., Putaux, JL., Heux, L., Dubreuil, F., & Rochas, C. The shape and size distribution of crystalline nanoparticles prepared by acid hydrolysis of native cellulose. Biomacromolecules 2007, 9, 57–65.
[9] Marchessault, RH., Morehead, FF., & Koch, MJ. Some hydrodynamic properties of neutral suspensions of cellulose crystallites as related to size and shape. J Colloid Sci 1961, 16, 327–344.
[10] Devi, RR., Dhar, P., Kalamdhad, A., & Katiyar, V. Fabrication of cellulose nanocrystals from agricultural compost. Compost Sci Util 2015, 23, 104–116.
[11] Teixeira, EDM., Pasquini, D., Curvelo, AA., Corradini, E., Belgacem, MN., & Dufresne, A. Cassava bagasse cellulose nanofibrils reinforced thermoplastic cassava starch. Carbohydr Polym 2009, 78, 422–431.
[12] Zhou, Y., Fuentes-Hernandez, C., Khan, TM., Liu, JC., Hsu, J., Shim, JW., & Kippelen, B. Recyclable organic solar cells on cellulose nanocrystal substrates. Sci Rep 2013, 3, 1536.
[13] Dhar, P., Tarafder, D., Kumar, A., & Katiyar, V. Effect of cellulose nanocrystal polymorphs on mechanical, barrier and thermal properties of poly (lactic acid) based bionanocomposites. RSC Adv 2015, 5, 60426–60440.
[14] Lin, N., & Dufresne, A. Nanocellulose in biomedicine: current status and future prospect. Eur Polym J 2014, 59, 302–325.
[15] Dhar, P., Bhardwaj, U., Kumar, A., & Katiyar, V. Cellulose nanocrystals: A potential nanofiller for food packaging applications. Food Addit Packag 2014, 1162, 197–239.
[16] Dhar, P., Bhardwaj, U., Kumar, A., & Katiyar, V. Poly (3-hydroxybutyrate)/cellulose nanocrystal films for food packaging applications: barrier and migration studies. Polym Eng Sci 2015, 55, 2388–2395.
[17] Zhao, LM., Shi, LE., Zhang, ZL., Chen, JM., Shi, DD., Yang, J., & Tang, ZX. Preparation and application of chitosan nanoparticles and nanofibers. Brazilian J Chem Eng 2011, 28, 353–362.
[18] Nemtsev, SV., Zueva, OY., Khismatullin, MR., Albulov, AI., & Varlamov, VP. Isolation of chitin and chitosan from honeybees. Appl Biochem Micro 2004, 40, 39–43.
[19] Paulino, AT., Simionato, JI., Garcia, JC., & Nozaki, J. Characterization of chitosan and chitin produced from silkworm crysalides. Carbohydr Polym 2006, 64, 98–103.
[20] Dash, M., Chiellini, F., Ottenbrite, RM., & Chiellini, E. Chitosan – A versatile semi-synthetic polymer in biomedical applications. Prog Polym Sci 2011, 36, 981–1014.
[21] Tang, ZX., Qian, JQ., & Shi, LE. Preparation of chitosan nanoparticles as carrier for immobilized enzyme. Appl Biochem Biotechnol 2007, 136, 77–96.
[22] Nair, LS., & Laurencin, CT. Biodegradable polymers as biomaterials. Prog Polym Sci 2007, 32, 762–798.

[23] Balázs, N., & Sipos, P. Limitations of pH-potentiometric titration for the determination of the degree of deacetylation of chitosan. Carbohydr Res 2007, 342, 124–130.
[24] Baskar, D., & Kumar, TS. Effect of deacetylation time on the preparation, properties and swelling behavior of chitosan films. Carbohydr Polym 2009, 78, 767–772.
[25] Feng, F., Liu, Y., Zhao, B., & Hu, K. Characterization of half N-acetylated chitosan powders and films. Procedia Eng 2012, 27, 718–732.
[26] Kaur, S., & Dhillon, GS. Recent trends in biological extraction of chitin from marine shell wastes: a review. Crit Rev Biotechnol 2015, 35, 44–61.
[27] Kim, TH., Jiang, HL., Jere, D., Park, IK., Cho, MH., Nah, JW., & Cho, CS. Chemical modification of chitosan as a gene carrier in vitro and in vivo. Prog Polym Sci 2007, 32, 726–753.
[28] Bhaskar, N., Suresh, PV., Sakhare, PZ., & Sachindra, NM. Shrimp biowaste fermentation with Pediococcus acidolactici CFR2182: optimization of fermentation conditions by response surface methodology and effect of optimized conditions on deproteination/demineralization and carotenoid recovery. Enzyme Microb Technol 2007, 40, 1427–1434.
[29] Yang, JK., Shih, L., Tzeng, YM., & Wang, SL. Production and purification of protease from a Bacillus subtilis that can deproteinize crustacean wastes☆. Enzyme Microb Technol 2000, 26, 406–413.
[30] Sashiwa, H., & Aiba, SI. Chemically modified chitin and chitosan as biomaterials. Prog Polym Sci 2004, 29, 887–908.
[31] Pillai, CKS., Paul, W., & Sharma, CP. Chitin and chitosan polymers: chemistry, solubility and fiber formation. Prog Polym Sci 2009, 34, 641–678.
[32] Pal, AK., & Katiyar, V. Non-isothermal degradation kinetics of PLA-chitosan films prepared by solution casting method. Asian Chitin J 2015, 11, 37–44.
[33] Wang, Z., Zheng, L., Li, C., Zhang, D., Xiao, Y., Guan, G., & Zhu, W. A novel and simple procedure to synthesize chitosan-graft-polycaprolactone in an ionic liquid. Carbohydr Polym 2013, 94, 505–510.
[34] Zhou, W., Wang, Y., Jian, J., & Song, S. Self-aggregated nanoparticles based on amphiphilic poly (lactic acid)-grafted-chitosan copolymer for ocular delivery of amphotericin B. Int J Nanomedicine 2013, 8, 3715.
[35] Yao, F., Liu, C., Chen, W., Bai, Y., Tang, Z., & Yao, K. Synthesis and characterization of chitosan grafted oligo (l-lactic acid). Macromol Biosci 2003, 3, 653–656.
[36] Pal, AK., & Katiyar, V. Nanoamphiphilic chitosan dispersed poly (lactic acid) bionanocomposite films with improved thermal, mechanical, and gas barrier properties. Biomacromolecules 2016, 17, 2603–2618.
[37] Peesan, M., Rujiravanit, R., & Supaphol, P. Electrospinning of hexanoyl chitosan/polylactide blends. J Biomater Sci Polym Ed 2006, 17, 547–565.
[38] Schreiber, SB., Bozell, JJ., Hayes, DG., & Zivanovic, S. Introduction of primary antioxidant activity to chitosan for application as a multifunctional food packaging material. Food Hydrocoll 2013, 33, 207–214.
[39] Soares, FC., Yamashita, F., Müller, CM., & Pires, AT. Thermoplastic starch/poly (lactic acid) sheets coated with cross-linked chitosan. Polym Test 2013, 32, 94–98.
[40] Vasile, C., Darie, RN., Cheaburu-Yilmaz, CN., Pricope, GM., Bračič, M., Pamfil, D., & Duraccio, D. Low density polyethylene–Chitosan composites. Compos B Eng 2013, 55, 314–323.
[41] Abdollahi, M., Rezaei, M., & Farzi, G. A novel active bionanocomposite film incorporating rosemary essential oil and nanoclay into chitosan. J Food Eng 2012, 111, 343–350.
[42] Ali, BH., Ziada, A., & Blunden, G. Biological effects of gum arabic: a review of some recent research. Food Chem Toxicol 2009, 47, 1–8.
[43] Akiyama, Y., Eda, S., & Katō, K. Gum arabic is a kind of arabinogalactan–Protein. Agric Biol Chem 1984, 48, 235–237.

[44] Dror, Y., Cohen, Y., & Yerushalmi-Rozen, R. Structure of gum arabic in aqueous solution. J Polym Sci B 2006, 44, 3265–3271.
[45] Mahdavi, SA., Jafari, SM., Assadpoor, E., & Dehnad, D. Microencapsulation optimization of natural anthocyanins with maltodextrin, gum Arabic and gelatin. Int J Biol Macromol 2016, 85, 379–385.
[46] García-Saldaña, JS., Campas-Baypoli, ON., López-Cervantes, J., Sánchez-Machado, DI., Cantú-Soto, EU., & Rodríguez-Ramírez, R. Microencapsulation of sulforaphane from broccoli seed extracts by gelatin/gum arabic and gelatin/pectin complexes. Food Chem 2016, 201, 94–100.
[47] Parija, S., Misra, M., & Mohanty, AK. Studies of natural gum adhesive extracts: an overview. J Macromol Sci C 2001, 41, 175–197.
[48] Shahid, M., Bukhari, SA., Gul, Y., Munir, H., Anjum, F., Zuber, M., & Zia, KM. Graft polymerization of guar gum with acryl amide irradiated by microwaves for colonic drug delivery. Int J Biol Macromol 2013, 62, 172–179.
[49] Abdel-Halim, ES., & Al-Deyab, SS. Hydrogel from crosslinked polyacrylamide/guar gum graft copolymer for sorption of hexavalent chromium ion. Carbohydr Polym 2011, 86, 1306–1312.
[50] Manna, PJ., Mitra, T., Pramanik, N., Kavitha, V., Gnanamani, A., & Kundu, PP. Potential use of curcumin loaded carboxymethylated guar gum grafted gelatin film for biomedical applications. Int J Biol Macromol 2015, 75, 437–446.
[51] Saurabh, CK., Gupta, S., Bahadur, J., Mazumder, S., Variyar, PS., & Sharma, A. Radiation dose dependent change in physiochemical, mechanical and barrier properties of guar gum based films. Carbohydr Polym 2013, 98, 1610–1617.
[52] Rao, MS., Kanatt, SR., Chawla, SP., & Sharma, A. Chitosan and guar gum composite films: preparation, physical, mechanical and antimicrobial properties. Carbohydr Polym 2010, 82, 1243–1247.
[53] Gupta, NR., Ghute, PP., & Badiger, MV. Synthesis and characterization of thermo-sensitive graft copolymer of carboxymethyl guar and poly (N-isopropylacrylamide). Carbohydr Polym 2011, 83, 74–80.
[54] Oprea, S. Effects of guar gum content on structure and properties of multi-crosslinked polyurethane composite films. Compos Part B: Eng 2013, 44, 76–83.
[55] Riggi, E., Santagata, G., & Malinconico, M. Bio-based and biodegradable plastics for use in crop production. Recent patents on food, nutrition & agriculture 2011, 3, 49–63.
[56] Banerjee, C., Ghosh, S., Sen, G., Mishra, S., Shukla, P., & Bandopadhyay, R. Study of algal biomass harvesting using cationic guar gum from the natural plant source as flocculant. Carbohydr Polym 2013, 92, 675–681.
[57] Soppirnath, KS., & Aminabhavi, TM. Water transport and drug release study from cross-linked polyacrylamide grafted guar gum hydrogel microspheres for the controlled release application. Eur J Pharm Biopharm 2002, 53, 87–98.
[58] Pandey, PK., Srivastava, A., Tripathy, J., & Behari, K. Graft copolymerization of acrylic acid onto guar gum initiated by vanadium (V)–Mercaptosuccinic acid redox pair. Carbohydr Polym 2006, 65, 414–420.
[59] Das, D., Ara, T., Dutta, S., & Mukherjee, A. New water resistant biomaterial biocide film based on guar gum. Bioresour Technol 2011, 102, 5878–5883.
[60] Wan, X., Li, Y., Wang, X., Chen, S., & Gu, X. Synthesis of cationic guar gum-graft-polyacrylamide at low temperature and its flocculating properties. Eur Polym J 2007, 43, 3655–3661.
[61] Garcıa-Ochoa, F., Santos, VE., Casas, JA., & Gomez, E. Xanthan gum: Production, recovery, and properties. Biotechnol Adv 2000, 18, 549–579.

[62] Talukdar, MM., & Kinget, R. Swelling and drug release behaviour of xanthan gum matrix tablets. Int J Pharm 1995, 120, 63–72.
[63] Lo, YM., Yang, ST., & Min, DB. Kinetic and feasibility studies of ultrafiltration of viscous xanthan gum fermentation broth. J Membrane Sci 1996, 117, 237–249.
[64] Bilanovic, D., Chang, FH., Isobaev, P., & Welle, P. Lactic acid and xanthan fermentations on an alternative potato residues media–Carbon source costs. Biomass Bioenergy 2011, 35, 2683–2689.
[65] Basavaraju, KC., Demappa, T., & Rai, SK. Miscibility studies of polysaccharide Xanthan gum and PEO (polyethylene oxide) in dilute solution. Carbohydr Polym 2007, 69, 462–466.
[66] Ibrahim, MS., Ibrahim, SM., & Farag, SA. Characterization, thermal, and mechanical behaviours of gamma irradiated gum arabic/poly vinyl alcohol polymer blends. Polym Plast Technol Eng 2007, 46, 1143–1149.
[67] Viljanmaa, M., Södergård, A., & Törmälä, P. Lactic acid based polymers as hot melt adhesives for packaging applications. Int J Adhes Adhes 2002, 22, 219–226.
[68] Altay, F., & Gunasekaran, S. Mechanical spectra and calorimetric evaluation of gelatin–Xanthan gum systems with high levels of co-solutes in the glassy state. Food Hydrocoll 2013, 30, 531–540.
[69] Veiga-Santos, P., Oliveira, LM., Cereda, MP., Alves, AJ., & Scamparini, ARP. Mechanical properties, hydrophilicity and water activity of starch-gum films: Effect of additives and deacetylated xanthan gum. Food Hydrocoll 2005, 19, 341–349.
[70] Aguirre-Cruz, A., Méndez-Montealvo, G., Solorza-Feria, J., & Bello-Pérez, LA. Effect of carboxymethylcellulose and xanthan gum on the thermal, functional and rheological properties of dried nixtamalised maize masa. Carbohydr Polym 2005, 62, 222–231.
[71] Jerez, A., Partal, P., Martínez, I., Gallegos, C., & Guerrero, A. Protein-based bioplastics: Effect of thermo-mechanical processing. Rheol Acta 2007, 46, 711–720.
[72] Perez, V., Felix, M., Romero, A., & Guerrero, A. Characterization of pea protein-based bioplastics processed by injection moulding. Food Bioprod Process 2016, 97, 100–108.
[73] Johansson, C., Bras, J., Mondragon, I., Nechita, P., Plackett, D., Simon, P., & Aucejo, S. Renewable fibers and bio-based materials for packaging applications–A review of recent developments. BioResources 2012, 7, 2506–2552.
[74] Sorrentino, A., Gorrasi, G., & Vittoria, V. Potential perspectives of bio-nanocomposites for food packaging applications. Trends Food Sci Technol 2007, 18, 84–95.
[75] Avella, M., De Vlieger, JJ., Errico, ME., Fischer, S., Vacca, P., & Volpe, MG. Biodegradable starch/clay nanocomposite films for food packaging applications. Food Chem 2005, 93, 467–474.
[76] Peelman, N., Ragaert, P., De Meulenaer, B., Adons, D., Peeters, R., Cardon, L., & Devlieghere, F. Application of bioplastics for food packaging. Trends Food Sci Technol 2013, 32, 128–141.
[77] Vollrath, F., & Porter, D. Silks as ancient models for modern polymers. Polymer 2009, 50, 5623–5632.
[78] Hardy, JG., Römer, LM., & Scheibel, TR. Polymeric materials based on silk proteins. Polymer 2008, 49, 4309–4327.
[79] Bini, E., Knight, DP., & Kaplan, DL. Mapping domain structures in silks from insects and spiders related to protein assembly. J Mol Biol 2004, 335, 27–40.
[80] Van Beek, JD., Hess, S., Vollrath, F., & Meier, BH. The molecular structure of spider dragline silk: Folding and orientation of the protein backbone. Proc Natl Acad Sci 2002, 99, 10266–10271.
[81] Gosline, JM., Guerette, PA., Ortlepp, CS., & Savage, KN. The mechanical design of spider silks: From fibroin sequence to mechanical function. J Exp Biol 1999, 202, 3295–3303.
[82] Hardy, JG., & Scheibel, TR. Composite materials based on silk proteins. Prog Polym Sci 2010, 35, 1093–1115.

[83] Padamwar, MN., & Pawar, AP. Silk sericin and its applications: a review. J Sci Ind Res 2004, 63, 323–329.
[84] Zhang, YQ. Applications of natural silk protein sericin in biomaterials. Biotechnol Adv 2002, 20, 91–100.
[85] Lawrence, BD., Wharram, S., Kluge, JA., Leisk, GG., Omenetto, FG., Rosenblatt, MI., & Kaplan, DL. Effect of hydration on silk film material properties. Macromol Biosci 2010, 10(4), 393–403.
[86] Vepari, C., & Kaplan, DL. Silk as a biomaterial. Prog Polym Sci 2007, 32(8–9), 991–1007.
[87] Barghout, JYJ., Thiel, BL., & Viney, C. Spider (Araneus diadematus) cocoon silk: A case of non-periodic lattice crystals with a twist?. Int J Biol Macromol 1999, 24, 211–217.
[88] Lee, KH., Baek, DH., Ki, CS., & Park, YH. Preparation and characterization of wet spun silk fibroin/poly (vinyl alcohol) blend filaments. Int J Biol Macromol 2007, 41, 168–172.
[89] Altman, GH., Horan, RL., Lu, HH., Moreau, J., Martin, I., Richmond, JC., & Kaplan, DL. Silk matrix for tissue engineered anterior cruciate ligaments. Biomaterials 2002, 23, 4131–4141.
[90] Wang, Y., Blasioli, DJ., Kim, HJ., Kim, HS., & Kaplan, DL. Cartilage tissue engineering with silk scaffolds and human articular chondrocytes. Biomaterials 2006, 27, 4434–4442.
[91] Min, BM., Lee, G., Kim, SH., Nam, YS., Lee, TS., & Park, WH. Electrospinning of silk fibroin nanofibers and its effect on the adhesion and spreading of normal human keratinocytes and fibroblasts in vitro. Biomaterials 2004, 25, 1289–1297.
[92] Jin, HJ., Fridrikh, SV., Rutledge, GC., & Kaplan, DL. Electrospinning Bombyx mori silk with poly (ethylene oxide). Biomacromolecules 2002, 3, 1233–1239.
[93] Koh, LD., Cheng, Y., Teng, CP., Khin, YW., Loh, XJ., Tee, SY., & Han, MY. Structures, mechanical properties and applications of silk fibroin materials. Prog Polym Sci 2015, 46, 86–110.
[94] Kesenci, K., Motta, A., Fambri, L., & Migliaresi, C. Poly (ε-caprolactone-co-D, L-lactide)/silk fibroin composite materials: Preparation and characterization. J Biomater Sci Polym Ed 2001, 12, 337–351.
[95] Zhou, S., Peng, H., Yu, X., Zheng, X., Cui, W., Zhang, Z., & Li, F. Preparation and characterization of a novel electrospun spider silk fibroin/poly (D, L-lactide) composite fiber. J Phys Chem B 2008, 112, 11209–11216.
[96] Arai, T., Wilson, DL., Kasai, N., Freddi, G., Hayasaka, S., & Tsukada, M. Preparation of silk fibroin and polyallylamine composites. J Appl Polym Sci 2002, 84, 1963–1970.
[97] Liu, Y., Shao, Z., Zhou, P., & Chen, X. Thermal and crystalline behaviour of silk fibroin/nylon 66 blend films. Polymer 2004, 45, 7705–7710.
[98] Freddi, G., Tsukada, M., & Beretta, S. Structure and physical properties of silk fibroin/polyacrylamide blend films. J Appl Polym Sci 1999, 71, 1563–1571.
[99] Jiang, C., Wang, X., Gunawidjaja, R., Lin, Y. H., Gupta, M. K., Kaplan, D. L., … Tsukruk, V. V. Mechanical properties of robust ultrathin silk fibroin films. Adv Funct Mater 2007, 17, 2229–2237.
[100] Tsuboi, Y., Ikejiri, T., Shiga, S., Yamada, K., & Itaya, A. Light can transform the secondary structure of silk protein. Appl Phys A 2001, 73, 637–640.
[101] Freddi, G., Romanò, M., Massafra, MR., & Tsukada, M. Silk fibroin/cellulose blend films: Preparation, structure, and physical properties. J Appl Polym Sci 1995, 56, 1537–1545.
[102] Wang, J., Shang, J., Ren, F., & Leng, X. Study of the physical properties of whey protein: Sericin protein-blended edible films. Eur Food Res Technol 2010, 231, 109–116.
[103] Li, M., Minoura, N., Dai, L., & Zhang, L. Preparation of Porous Poly (vinyl alcohol)-Silk Fibroin (PVA/SF) Blend Membranes. Macromol Mater Eng 2001, 286, 529–533.

[104] Kweon, HY., Park, SH., Yeo, JH., Lee, YW., & Cho, CS. Preparation of semi-interpenetrating polymer networks composed of silk fibroin and poly (ethylene glycol) macromer. J Appl Polym Sci 2001, 80, 1848–1853.
[105] Kim, UJ., Park, J., Li, C., Jin, HJ., Valluzzi, R., & Kaplan, DL. Structure and properties of silk hydrogels. Biomacromolecules 2004, 5, 786–792.
[106] Gil, ES., Spontak, RJ., & Hudson, SM. Effect of β-sheet crystals on the thermal and rheological behavior of protein-based hydrogels derived from gelatin and silk fibroin. Macromol Biosci 2005, 5, 702–709.
[107] Hu, K., Lv, Q., Cui, FZ., Xu, L., Jiao, YP., Wang, Y., & Huang, LY. A novel poly (L-lactide)(PLLA)/fibroin hybrid scaffold to promote hepatocyte viability and decrease macrophage responses. J Bioact Compat Polym 2007, 22, 395–410.
[108] Chen, G., Zhou, P., Mei, N., Chen, X., Shao, Z., Pan, L., & Wu, C. Silk fibroin modified porous poly (ε-caprolactone) scaffold for human fibroblast culture in vitro. J Mater Sci Mater Med 2004, 15, 671–677.
[109] Swinerd, VM., Collins, AM., Skaer, NJ., Gheysens, T., & Mann, S. Silk inverse opals from template-directed β-sheet transformation of regenerated silk fibroin. Soft Matter 2007, 3, 1377–1380.
[110] Nazarov, R., Jin, HJ., & Kaplan, DL. Porous 3-D scaffolds from regenerated silk fibroin. Biomacromolecules 2004, 5, 718–726.
[111] Demura, M., & Asakura, T. Porous membrane of Bombyx mori silk fibroin: Structure characterization, physical properties and application to glucose oxidase immobilization. J Membrane Sci 1991, 59, 39–52.
[112] Arai, T., Freddi, G., Innocenti, R., & Tsukada, M. Biodegradation of Bombyx mori silk fibroin fibers and films. J Appl Polym Sci 2004, 91, 2383–2390.
[113] Kasoju, N., & Bora, U. Silk fibroin in tissue engineering. Adv Healthc Mater 2012, 1, 393–412.
[114] Postlethwait, RW., Willigan, DA., & Ulin, AW. Human tissue reaction to sutures. Ann Surg 1975, 181, 144.
[115] Yucel, T., Cebe, P., & Kaplan, DL. Vortex-induced injectable silk fibroin hydrogels. Biophys J 2009, 97, 2044–2050.
[116] Uebersax, L., Mattotti, M., Papaloïzos, M., Merkle, HP., Gander, B., & Meinel, L. Silk fibroin matrices for the controlled release of nerve growth factor (NGF). Biomaterials 2007, 28, 4449–4460.
[117] Lu, S., Wang, X., Lu, Q., Hu, X., Uppal, N., Omenetto, FG., & Kaplan, DL. Stabilization of enzymes in silk films. Biomacromolecules 2009, 10, 1032–1042.
[118] Kim, DH., Viventi, J., Amsden, JJ., Xiao, J., Vigeland, L., Kim, YS., & Kaplan, DL. Dissolvable films of silk fibroin for ultrathin conformal bio-integrated electronics. Nat Mater 2010, 9, 511.
[119] Tsioris, K., Raja, WK., Pritchard, EM., Panilaitis, B., Kaplan, DL., & Omenetto, FG. Fabrication of silk microneedles for controlled-release drug delivery. Adv Funct Mater 2012, 22, 330–335.
[120] Tao, H., Kainerstorfer, JM., Siebert, SM., Pritchard, EM., Sassaroli, A., Panilaitis, BJ., & Kaplan, DL. Implantable, multifunctional, bioresorbable optics. Proc Natl Acad Sci 2012, 109, 19584–19589.

Arvind Gupta, Medha Mili, Tabli Ghosh and Vimal Katiyar

4 Polylactic acid for food packaging applications

4.1 Introduction

The renewable precursor-based chemically synthesized polymers can be classified as bio-derived polymers. Among all chemically synthesized bioplastics, polylactic acid(s) (PLA) has acquired an extensive multiplicity in replacing petrochemical-based conventional polymers, and indeed have a wide commercial availability in the food packaging sector. PLA is a thermoplastic and comes in the category of aliphatic polyester. It can be produced by polymerization of lactic acid (LA), which can be derived from agricultural feedstock. This chapter mainly provides an insight about different prospects of PLA. Its properties, limitations, synthesis mechanisms and various research findings in relation with its applications for food packaging have been outlined.

4.2 PLA

PLA is classified as a biodegradable aliphatic polyester and is mainly synthesized using precursors obtained from natural resources [1]. Nowadays, PLA is commercially viable and has a great potential for being used in an enormous number of applications [2]. Moreover, due to its nontoxic nature [3], PLA is considered as a favorable material in food and beverage industries, and other consumer goods applications. The easy processing of PLA by conventional methods such as thermoforming and injection blow moulding is due to its greater thermal processability in comparison to other bioplastics, which makes PLA superior to other conventional thermoplastics. These properties of PLA have greatly contributed toward making this polymer a potential candidate for food packaging applications. The basic building block of PLA is LA (a simple hydroxycarboxylic acid) and is widely distributed in nature [4]. The production of LA can easily be done from carbohydrate-based feedstock such as corn and tapioca starch. The two optically active isomers of LA include L(+) LA and D(+) LA, and lactide is another intermediate monomer for PLA synthesis, which is primarily produced through backbiting reaction of OH end groups in oligomeric PLA [5]. Figure 4.1 represents L-lactide, D-lactide and meso-lactide.

https://doi.org/10.1515/9783110648034-004

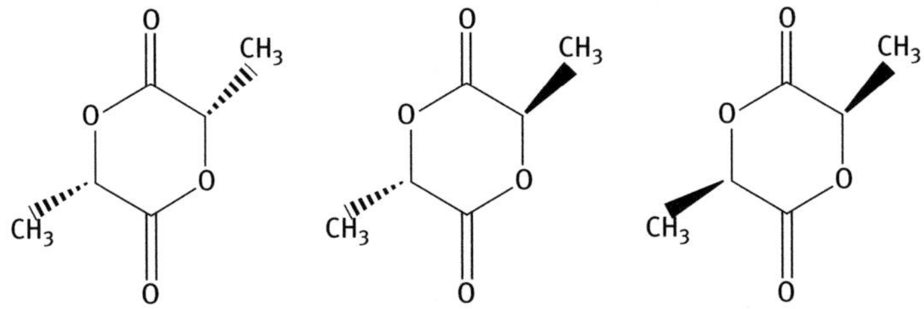

Figure 4.1: Structures of three types of lactide: (a) L-lactide, (b) meso-lactide and (c) D-lactide.

4.3 Synthesis of PLA

The synthesis of PLA mainly occurs through three synthetic routes such as (i) direct polycondensation of LA, (ii) azeotropic dehydrative condensation and (iii) ring-opening polymerization (ROP) [6].

Polycondensation reaction is the simplest and least expensive route to synthesize PLA directly from the LA monomer, through a reversible step-growth mechanism. In this approach, the removal of byproduct such as water limits the ultimate molecular weight (MW) achievable, resulting in the synthesis of a low MW, brittle, glassy polymer, which makes the most part of the material unusable from commercial point of view [7]. Thus, to make it commercially feasible, various external coupling and chain-extending candidates are used to enhance the MW of the polymer, where the main disadvantages associated with the final polymers include the presence of toxic and nonbiodegradable chain-extending agents [2] and hence limits the use of PLA in medical and food packaging sectors.

Subsequently, in order to overcome the above limitations, PLA synthesis is carried out using catalyst and organic solvents via azeotropic dehydrative condensation process without using any chain-extending agents. In this process, water is removed azeotropically as a byproduct, whereas solvent is dried and recycled back into the reaction [8]. This technique yields high-MW PLA, but leaves behind residual catalysts that can cause many problems during further processing and in various applications in medical or food packaging sectors.

Another, most preferable technique and commonly used route to achieve high-MW PLA is the ROP [9] of lactide. Figure 4.2 shows the direct condensation polymerization and ROP synthesis of PLA. This process involves mainly three steps: (i) polycondensation of LA to obtain a low MW prepolymer, (ii) depolymerization of the low MW prepolymer to form a cyclic ester known as lactide, and finally (iii) ROP of the lactide occurs to synthesize PLA with controlled MW in the presence of suitable types of organometallic catalysts [10, 11], which can be carried out in melt [12], bulk [13–15] or in solution [16] or by cationic, anionic and coordination-insertion mechanisms. Among the various types of organometallic catalysts, stannous octoate is the most widely used type because of its high catalytic activity and permitted use from the US Food and Drug Administration (FDA) [2], which makes it ideal for utilization in food packaging applications. But the main limitation associated with the organometallic catalysts is their difficulty in the removal of the metal traces from the final polymeric product which may be toxic, and hence, pose difficulty in food-packaging application. Thus, nowadays, the main focus of researchers is on the utilization of various metal-free/organocatalysts to synthesize metal-free PLA to avoid toxicity in food packaging applications.

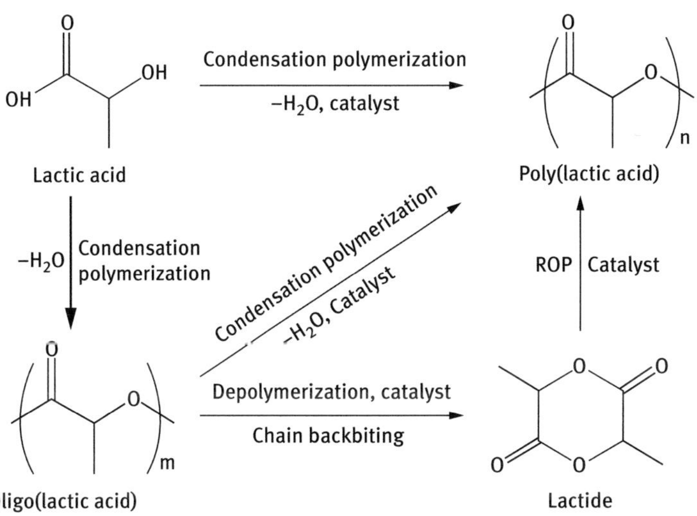

Figure 4.2: Schematic of the different routes for the synthesis of PLA.

4.4 Properties of PLA

The high-MW PLA is a glossy, colorless, stiff thermoplastic polymer, which exhibits tensile strength comparable to other commercially available polymers. Polylactides that exist in three isomeric forms mainly determine the final properties of the polymer. Among the polylactides available, poly(L-lactide) (PLLA) and poly(D-lactide) (PDLA) are crystalline polymers with a melting point around 180 °C. However, racemic lactide (50% D and 50% L) mixture gives an amorphous polymer, poly(DL-lactide), with a glass transition temperature of 50–57°C. Also, it has been clearly explained in reported literature that the properties of polylactides to a great extent, are based on the isomeric ratio and its distribution along with the MW of the polymer. The investigation of rheological properties is also an important parameter for the evaluation of the nature of thermoplastics in order to have their effective processing. PLA is found to be a pseudoplastic, non-Newtonian fluid, which behaves as a classic flexible-chain polymer across all optical compositions above its melting point [17]. In addition, for acting as a proper packaging material, the mechanical and barrier properties of PLA should be studied extensively for its wide application. PLA is found to be a brittle material having considerable mechanical properties such as tensile strength of 32 MPa, high Young's modulus value of 2.3 GPa and a low percentage elongation of break of about 5% as obtained from mechanical analysis [18]. Under the same testing conditions, the tensile strength values obtained for poly(L-lactide) are similar to polystyrene (PS) but lower than polyethylene terephthalate (PET). The melting temperature (T_m) and glass transition temperature (T_g) of PLA are significantly lower than PET and PS, which improve heat sealing and thermal processing capabilities of PLA. Also, focussing on food packaging applications, it is necessary to have a clear understanding of the barrier properties of PLA which are found to be higher than polyethylene and polypropylene, similar to PS but lower than those of PET [19]. In general, the crystallinity of PLA strongly affects the barrier properties of the polymer, whereas the decrease in crystallinity was a negative aspect in terms of the mechanical and barrier properties. These issues have emerged as a research area for the further improvement of the properties of PLA. Thus, PLA has been experimented for developing various nanocomposites in order to improve its various mechanical and thermal properties. Various types of nanofillers have been incorporated in order to improve the crystallinity, thermal, barrier and mechanical properties of neat PLA and its conventional composites, which results in the increased applications of PLA [20–22]. With the above-mentioned advantages, PLA has some limitations such as low heat stability and low barrier capability. Due to these limitations, applications of PLA in various fields such as biomedical and food packaging applications have been limited up to some extent. For this, several research groups are currently working to increase the properties of PLA by using different techniques such as reinforced fillers, blending with other polymers and so on.

4.5 Stereocomplex PLA

PLA has some limitations such as low heat stability (thermomechanical stability) and high gas permeability. Due to these limitations, PLA has limited application in biomedical, not in engineering. Hence, modifications of PLA properties can be done by reinforcing fillers, blending with other polymers and so on. Some of the studies on PLA property enhancement are discussed below.

In 1987, Ikada et al. [23] had reported the formation of stereocomplex in PLA, which had the melting temperature of ~50 °C higher than normal PLA which opened up a new window in the direction to develop the biodegradable polymer which may have the capability to replace the petroleum-based polymer for engineering applications also. Stereocomplexation in PLA is a crystalline arrangement of two different enantiomeric PLA chains (PLLA and PDLA mix with 1:1 ratio). The main limitation in the synthesis of stereocomplex PLA (sPLA) is the formation of homocrystal with stereocomplex crystallites. Also, it forms only with relatively low MW (<100 KDa) PLA, which affect the mechanical properties of final product. Furthermore, different techniques are being developed by the researchers to improve the above properties. A polymer stereocomplex is mainly a stereoselective interaction between two complementing stereoregular polymers that interlock and form a new composite, demonstrating altered physical properties in comparison to the parent polymers [24]. Tsuji [25] had studied the interaction between polymers having different tacticities or configurations, which prevailed over polymers with the same tacticity or configuration and resulted in a stereoselective association of the former polymer. Such association is described as stereocomplexation or stereocomplex formation. Stereocomplex can form between isotactic and syndiotactic or optically active or opposite enantiomeric polymers.

Ikada et al. [23] reported the synthesis of sPLA by solution mixing of PLLA and PDLA. They dissolved PLLA or PLDA separately into methylene chloride and then mixed to each other. After precipitation into methanol, stereocomplex sPLA was formed. The melting point of synthesized sPLA was enhanced to 50 °C than normal PLA. The different crystalline structure also proved the sPLA formation. They suggested that the stereocomplex was formed through Van der Waals forces such as dipole–dipole interactions between the two different helical chains in solution, where molecular motion is sufficiently great.

Further, several methods have been developed by different scientific groups for the formation of stereocomplex in PLA. Methods such as solution casting, melt blending, precipitation and stereoblock are used to develop stereocomplexation in PLA. Furthermore, due to interaction between enantiomeric PLA, that is, PLLA and PDLA, it forms different crystalline structure from enantiomeric pure PLA. So, for the formation of stereocomplex, molecular interaction between PLLA and PDLA is the main factor. This interaction could be done by mixing low MW PLLA and PDLA by solution cast or melt blending, which in contrast leads to inferior mechanical properties of end product. Based on this discussion, it is very important to have high-MW sPLA for engineering applications. Yui et al. [26] reported the formation of stereocomplex crystallites in the case of block copolymers of L and D-lactide. They assumed that the degree of stereocomplex formation could be tuned by the composition of the block copolymers. They synthesized the block copolymer by solution polymerization of L-lactide in toluene and subsequent addition of D-lactide, where the melting point was found at 205 °C.

Stereocomplexation can be examined by several methods such as nuclear magnetic resonance (NMR), differential scanning calorimetry (DSC) and X-ray diffraction (XRD). It was also found that the complex of PLLA and PDLA was not soluble in the solvents, which can dissolve the enantiomeric PLA. However, hexafluoroisopropanol can be used to dissolve the sPLA. So this is also a method to confirm the formation of sPLA.

Due to different crystallite formation in sPLA, it is very easy to analyze using XRD technique. The enantiomeric pure PLA exists in several polymorphs such as α, β and γ forms. Among these, alpha form is found to be the most stable form of PLA, which crystallizes in the pseudo-orthorhombic unit cell with dimensions $a = 1.07$ nm, $b = 0.595$ nm and $c = 2.78$ nm formed by two 10_3 helices. Whereas sPLA takes 3_1 helical confirmations by parallel packing of L- and D-lactide segments of chains in a triclinic unit cell of dimensions $a = b = 0.916$ nm, $c = 0.870$ nm, $\alpha = \beta = \gamma = 109.2°$. The XRD pattern for the sPLA had peaks at 2θ values of 11.9°, 20.9° and 23.9°, which corresponded to the (110), (300/030) and (220) planes of stereocomplex crystals [25]. The peaks at 16.7° and 19° corresponded to (110/200) and (203) planes of α-form of homocrystals of PLA. Similarly, using DSC, the content of stereocomplex crystallite could be quantified easily because the sPLA always had higher melting temperature than pure PLA.

Several companies commissioned industrial-level plants for the production of PLA and sPLA. Sulzer technology, Corbian, Purac and Nature works LLC as Ingeo trade names are the big companies in the field of production of PLA.

4.5.1 Use of a third component to improve the stereocomplexation in PLA

Several research groups are trying to improve the content of stereocomplex crystallites in PLA by addition of a third component other than PLLA and PDLA. Li and Woo [27] found that the addition of poly(4-vinyl phenol) (PVPh) in the PLLA/PDLA blend accelerated the crystallization and reduced the size of stereocomplex spherulites. Addition of low content of PVPh in the blend of PLLA/PDLA also accelerated the stereocomplexation of blend. Chang and Woo [28] had studied the effect of poly (3-hydroxybutyrate) (PHB) on the crystallization behavior of sPLA. However, the negative effect of addition of PHB in sPLA was observed. Sun and He [29] grafted GO with PDLA and subsequently blended with PLLA, which led to sPLA formation. It was found that incorporation of GO nanofiller enhanced the crystallinity and increased the fraction of stereocomplex crystallites content in the composite, whereas higher GO loading could act as blocking sites and restrict the polymer chain mobility which slows down the growth of the spherulites. Quan et al. [30] added carbon nanotubes (CNT) in the sPLA matrix using direct melt mixing process, which influenced the crystallinity of the composite due to its nucleating effect that enhanced up to 62 wt% only by addition of 1 wt% along with the improved electrical properties. Effect of dispersion of layered silicate and its influence on the stereocomplexation of PLA was also reported in the literature; however, mixed crystal growth was also obtained in the resulting nanocomposites [31]. Bao et al. [32] studied the effect of polyethylene glycol (PEG) on the formation of stereocomplexation in PLA. They found an increase in stereocomplex crystallites with increase in PEG content. The T_g of the material decreased from 62 to 37 °C due to the plasticizing effect of the PEG. It was also noticed that low MW PEG could act as an effective additive for the formation of stereocomplex crystallites. The low MW PEG increased the segmental movement of the polymeric chains, which led to the formation of sPLA crystallites. With the same aim to improve the stereocomplexation of PLA, Purnama and Kim [33] synthesized the sPLA using cellulose nanowhiskers under supercritical medium, where single sPLA stereocomplex crystallites having melting point of more than 200 °C were isolated. Further, zinc phenylphosphonate as a nucleation agent in the formation of sPLA was successfully demonstrated [34] with improved crystallization rate, which selectively promoted the formation of stereocomplex crystallites.

Table 4.1 lists the fillers that affect the formation of sPLA and connect the enantiomeric PLA MW with stereocomplexation of PLA that in principle affect the properties of end product. Moreover, reported research shows that the solution casting method can be a technique for the proof of concept but cannot be a commercial level method. On the other hand, melt processing of polymers is found to be an industrial method for large-scale production. So based on this discussion, there is a great requirement for the development of the processing method, which should enhance the stereocomplexation in PLA suitable for large-scale production.

4.5.2 Stereocomplex PLA for food packaging

PLA is available in the market and increased in food industries. For example, trays prepared by thermoforming technique are used for the packaging of meats. These packaging materials have the shortcomings in holding hot materials, such as hot tea or coffee. The glass transition temperature and heat deflection temperature of PLA is found to be in the range of 55–65 and 60–75 °C, respectively. In addition, the property of relatively low heat resistance of PLA can cause "blocking" in the stacked articles during transportation. In "blocking" phenomenon, the article can stick together during transportation due to low heat deflection and glass transition temperature. In contrast, sPLA can be a potential candidate for the packaging of the hot food and cups for hot tea or any liquid food. Bor et al. [35, 36] studied the effect of different food stimulant on sPLA and found that the release of oligomeric chains and cyclic LA is somewhat smaller than the enantiomeric pure PLA. They have also studied the effect of stereocomplexation on the stability migration

Table 4.1: Representative list of fillers affecting stereocomplexation and properties of PLA.

Sl. No.	Filler	Filler content (wt%)	Molecular weight (PLLA/PDLA)	Stereocomplex melting temperature (°C)	Stereocomplex crystallinity (%)	Reference
01	Poly-(*p*-vinyl phenol) (PVPh)[f]	10[c]	152 kDa/ 124 kDa (Mw)	220	Not shown	[6]
02	PDLA-grafted graphene oxide[g]	0.1[c]	130 kDa/ 12 kDa (Mn)	210	7.7[b]	[8]
03	Layered silicate[f]	10[c]	22 kDa/ 26.3 kDa (Mw)	217	19.4[a]	[10]
04	Poly[(*R*)-3-hydroxybutyrate][f]	30[c]	11 kDa/ 124 kDa (Mw)	221	Not shown	[7]
05	Multiwalled carbon nanotubes[f]	1.0[d]	105 kDa/not shown (Mw)	206	64.6[b]	[9]
06	Poly-(ethylene glycol)[f]	10[c]	210 kDa/ 100 kDa (Mw)	210	17.3[a]	[11]
07	Cellulose nanocrystals[f]	5[e]	130 kDa/ 150 kDa (Mw)	230	Not shown	[12]
08	Zinc phenylphosphonate[f]	0.5[c]	143 kDa/ 191 kDa (Mw)	230	17[a]	[13]

[a]Determined using DSC.
[b]Determined using XRD.
[c]Solvent cast method.
[d]Melt mixing.
[e]Supercritical fluid technology.
[f]PLLA:PDLA ratio = 50:50.
[g]PLLA:PDLA ratio 20:80.

resistance of the PLA in microwave heating with different food stimulants such as ethanol and water. They found that due to the less amorphous phase present in the sPLA, the amount of migrant such as linear oligomeric chain or cyclic LA was found in lesser amount in 10% ethanol than PLLA. The mass loss during microwave heating in water or 10% ethanol was less in case of stereocomplex than normal PLLA. This makes sPLA a potential candidate for application in food packaging.

4.6 Case study

4.6.1 Effect of incorporating LA oligomer-g-chitosan nanofiller in PLA films

Pal and Katiyar [37] have developed bionanocomposite films incorporating an amphiphilic molecule, that is, OLLA-*g*-CH as an effective nanofiller in PLA films. The research work mainly focused on improved thermal, mechanical and gas barrier properties of PLA/OLLA-*g*-CH composite films for acting as a perfect food-packaging material. In the study, OLLA-*g*-CH was synthesized using in situ condensation polymerization technique. Generally, grafting of polymers is a practice for making segmented copolymers, where branched polymers (OLLA) are tagged over backbone of other polymers (CH). Here, grafting between the two molecules was formed through amide–ester bonds, which worked as a surfactant having CH as hydrophilic head, and OLLA as hydrophobic tail. The characterization techniques such as transmission electron microscopy (TEM) and Fourier transform infrared spectroscopy (FTIR) confirm the grafting of OLLA with CH molecule. Figure 4.3a shows TEM analysis of OLLA-*g*-CH (30%) bionanocomposite, which reveals the distribution of OLLA-*g*-CH molecules (average length ~4–6 nm), and their spherical micelles formation. In addition, XRD analysis investigated about amorphous and orthorhombic crystal structures of CS and PLA, respectively. The nanofiller was found to have no effect on the ordered structure of PLA matrix in PLA/OLLA-*g*-CH. Polarized optical microscopy (POM) analysis demonstrated the improved spherulitic growth of PLA at 110 °C with the addition of OLLA-*g*-CH as shown in Figure 4.3b. The spherulite size of PLA was observed to increase with the increase in time as well as filler loading. The developed PLA-based bionanocomposite films

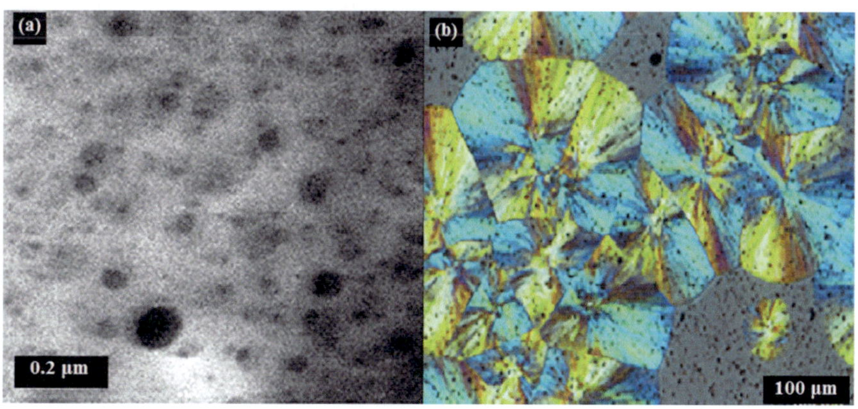

Figure 4.3: (a) TEM analysis of OLLA-*g*-CH bionanocomposite and (b) POM image of PLA/OLLA-*g*-CS (3%) at 110 °C. Adapted from Pal and Katiyar [37].

having uniform dispersion of OLLA-*g*-CH molecules (size range ~20–150 nm) were found to offer improved thermal, mechanical and gas barrier properties. The thermal properties of the prepared films were tested in terms of glass transition temperature (T_g), melting temperature (T_m) and heat capacity (C_p). With increase in filler loading, a decrease in T_g and C_p values was observed up to 18 °C and 0.4 J/g K, respectively. Furthermore, a noticeable decrease in melting temperature was also observed due to the plasticizing effect of OLLA-*g*-CH. The ultimate tensile strength of bionanocomposite films was found to reduce by ~18.4% in comparison with PLA films. Besides, a severe increment in percentage elongation of PLA/OLLA-*g*-CH films was detected due to the increase in plasticizing effect. Young's modulus of bionanocomposite film was found to decrease from 542 ± 133 to 298 ± 3 1 MPa with increase in filler concentration (up to 5%). Another crucial factor of packaging material, that is, gas barrier properties of prepared films were also improved due to the addition of filler material. As well known, CS molecules are hydrophilic in nature, whereas OLLA molecules are hydrophobic in nature, so drafting of OLLA with CS molecules tuned to overcome the shortcomings of CS molecules as filler materials in PLA film. Studies on oxygen transmission rate (OTR) provide the information about gas barrier properties of prepared bionanocomposite films. The investigation showed that there is ~10-fold decrease in OTR of bionanocomposite films as compared to PLA film. With increase in filler loading, both solubility and diffusivity of oxygen across the film were found to decrease the OTR values, whereas temperature had a negative impact on the OTR values and OTR was found to increase with an increase in temperature. The bionanocomposite films have shown an improved oxygen barrier properties by decreasing oxygen permeability (~10-fold). Basically, permeability is a function of solubility and diffusivity. Hence, the improved oxygen barrier results were observed due to the reduction in solubility of oxygen molecules across PLA films with the help of OLLA-*g*-CH molecules. Based on the above discussion, the developed PLA/OLLA-*g*-CH bionanocomposite films would outgrow an application toward food packaging due to its improved thermal, mechanical and gas barrier properties. Moreover, the newly developed films are environment friendly and posses no harmful effect to the surroundings.

4.6.2 Effect of incorporating LA-g-gum arabic nanofiller in PLA films

Tripathi and Katiyar [38] had developed a bionanocomposite film incorporating a nanofiller material, functionalized gum arabic (LA-*g*-GA) in PLA films. In their research and findings, microwave-synthesized LA was grafted with GA molecules by polycondensation reaction. The grafting of selected molecules provides the benefits of converting hydrophilic GA molecules to hydrophobic blend (LA-*g*-GA), and when incorporated in PLA films improves its properties (thermal, mechanical and gas barrier properties) for acting as a proper packaging material. The grafting of molecules was confirmed through ^{1}H-NMR analysis of LA-*g*-GA. Furthermore, TEM analysis showed the homogeneous dispersion of nanofillers (LA-*g*-GA) in PLA films (Figure 4.4). Studies on developed film [PLA/(LA-*g*-GA)] properties showed that there was a significant improvement in thermal, mechanical and gas barrier properties of PLA films. The thermal stability of the PLA/LA-*g*-GA films was decreased due to nanofiller materials with an increase in filler material. The observed T_g values were 64.8, 46.4, 41.8 and 41.0 °C for PLA, PLA/LA-*g*-GA (3 wt% loading), PLA/LA-*g*-GA (5 wt% loading) and PLA/LA-*g*-GA (10% wt loading), respectively. The improved mechanical properties of the developed film were determined in terms of tensile strength, elongation at break and Young's modulus. In comparison with neat PLA, the tensile strength was reduced by 14% and the elongation at break was enhanced by 16% for PLA/LA-*g*-GA 3% loading material (3%). Moreover, Young's modulus was found to decrease by 28.2%, 32.8% and 40.7% for 3%, 5% and 10% loading, respectively. In case of gas barrier properties, the percent reduction of OTR × *e* was found to be 88.3%, 90.3% and 89.1% for 3%, 5% and 10% filler loading in PLA film. The result was obtained due to the reduced oxygen solubility in the PLA film due to the filler material in the PLA film. On the other hand, water vapour transmission rate (WVTR) was also reduced by 7% and 27% for filler loading of 3% and 5%, respectively. In contrast, an abrupt rise was observed in WVTR for filler material (10%) films. So, the used filler materials (a renewable and environment-friendly source) were tuned to overcome the limitations of PLA films and have a wide application as biobased food-packaging material.

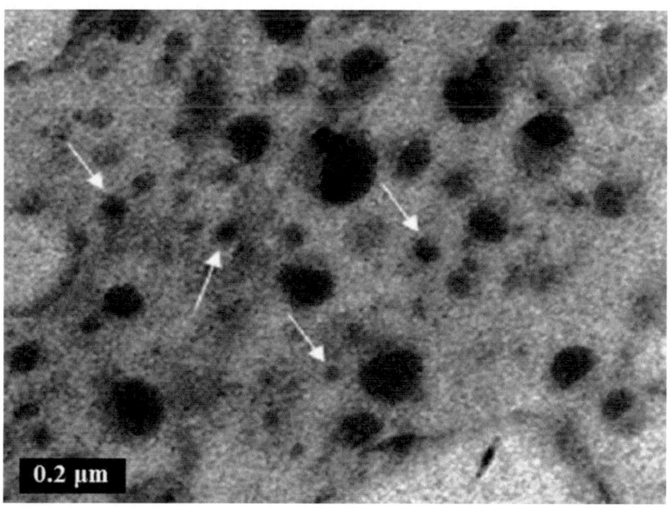

Figure 4.4: TEM images of PLA/(LA-*g*-GA) (5%). Adapted from Tripathi and Katiyar [38].

4.6.3 Effect of polymorphic cellulose nanocrystal nanofillers on the properties of PLA-based nanocomposite films

Dhar et al. [22] had studied the effect of incorporating CNC polymorphs (CNC I, CNC II and CNC: I→II from CNC I) on the properties of PLA films [22]. CNC polymorphs were extracted from bamboo stems as cellulosic precursor, through alkali treatments followed by acid hydrolysis approach. In their study, the effect of CNC polymorphs on various properties such as thermal, mechanical and gas barrier behavior of the biocomposite films were studied. The transformation of the cellulose I to cellulose II was confirmed through the XRD studies, which showed the transformation of the single peak in CNC I at 22.6° to splitting of two peaks in CNC II at 11.8° and 19.8°, respectively. In their study, evaluation of the physicochemical properties through FTIR analysis gave the evidence of transformation of CNC I to CNC II polymorphs, which was identified through small changes in their torsional angles between glycosidic linkages between β(1→4)-D-glucose unit. Besides, FTIR analysis illustrated the discrete separations among all the polymorphs of CNC, where board band of –OH bonds stretched in the range of 3,700–3,000 cm^{-1} IR spectrum. The surface morphology of polymorphs of CNC was characterized by AFM and FESEM micrographs (Figure 4.5a). The average width and height of the polymorphs were determined by AFM and FESEM, respectively. The average widths of the polymorphs CNC I, CNC II and CNC: I →II were 19 ± 5, 34 ± 8 and 100 ± 30 nm, respectively. The average lengths of CNC I, CNC II and CNC: I→II were 610 ± 55, 824 ± 60 and 583 ± 40 nm, respectively. Figure 4.5a represents the FESEM and AFM micrographs of three polymorphs of CNC. Moreover, fractural behavior of the tensile samples was studied by FESEM analysis, where alteration of samples to white color occurs by the application of stress, which was possibly due to strain-induced crystallization in the samples (Figure 4.5c). Studies on thermal, mechanical and barrier properties of prepared bionanocomposite film showed rigorous improvements in their properties due to variation in the inherent properties of the CNC polymorphs nanofiller. The Young's modulus of fabricated films significantly improved by 72% for CNC II and CNC I→II. In contrast, a significant decrease in elongation at break for the polymorphs CNC II and CNC I→II was observed in comparison to CNC I, which was due to their difference in hydroxyl functionality and degree of hydrogen bonding (Figure 4.5b). The reinforcement effect of the different CNC polymorphs was quantified using different mathematical modeling (Halpin–Kardos, Ouali and Cox–Krenchel models) of the prepared composite film which showed good agreement with experimental values. Thermogravimetric analysis stated that the thermal stability and degradation rates of PLA film could be tuned through incorporation of CNC polymorphs. Studies on barrier properties showed CNC II and CNC I →II, which introduces higher reduction in WVTR than CNC I. The WVTR was found to be reduced by ~38%, 31% and 40% for CNC I (1 wt% loading), CNC II (1 wt% loading) and CNC I→II (1 wt% loading), respectively. OTR of the films were reduced by ~69, 64, for CNC I (1 wt% loading), CNC II

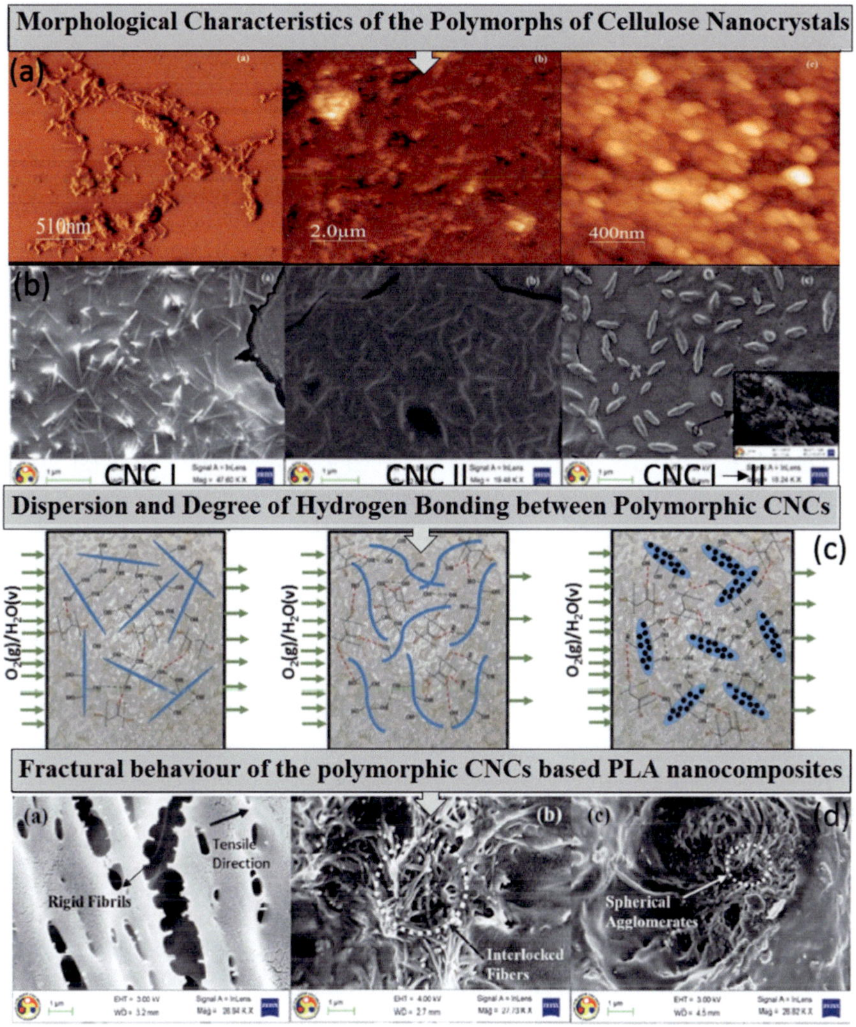

Figure 4.5: (a) AFM and FESEM micrographs of CNC polymorphs (a) CNC I, (b) CNC II, and (c) CNC I→II; (b) degree of the intra- or intermolecular hydrogen bonding between the different polymorphs of CNC and the PLA matrix; and (c) fractural behavior of PLA/CNC I, PLA/CNC II and PLA/CNC I→II nanocomposites, respectively.

(1 wt% loading), respectively. In their study it was observed that the barrier properties were unaffected with an increase in loadings of CNC I polymorphs to 3 wt% loading. For CNC II (3 wt% loading) and CNC I→II (3 wt% loading), the WVTR was reduced by ~51% and 48%, respectively. The OTR was reduced for CNC II (3 wt% loading) and CNC I→II (2 wt% loading) by ~80% and 63%, respectively. So, their study signified

that proper selection of CNC polymorphs followed by its dispersion in PLA matrix leads to the formation of biobased packaging films with tuneable properties depending on the targeted applications.

4.6.4 Effect of incorporating sucrose palmitate nanofillers in PLA films

Valapa et al. [39] studied the nonisothermal crystallization kinetics of sucrose palmitate (SP)-incorporated PLA bionanocomposite films [4]. In their study, SP, which is a nontoxic food flavoring agent, was incorporated in PLA films through the process of solution casting. The findings of the thermal properties and nonisothermal crystallization kinetics of the developed bionanocomposite film were carried out by using DSC under heating rates 2.5, 5, 7.5 and 10 °C/min. The observed T_g values were 61, 60, 60, 60 °C for the neat PLA, PLA/(SP 1 wt%), PLA/(SP 3 wt%) and PLA/(SP 5 wt%), respectively. Furthermore, a significant increase in cold crystallization temperature was observed with increase in loading wt%, which was due to the reduction in diffusion rate by SP nanofillers. The observed T_{cc} values from DSC thermogram obtained were 110, 117, 125 and 125 °C for PLA, PLA/(SP 1 wt%), PLA/(SP 3 wt%) and PLA/(SP 5 wt%), respectively. Besides, addition of filler material shifts the melting temperature of the PLA film, which was 149, 150, 152 and 152 °C for PLA, PLA/(SP 1 wt%), PLA/(SP 3 wt%) and PLA/(SP 5 wt%), respectively. Studies on the nonisothermal crystallization kinetics of prepared bionanocomposite film showed that for neat PLA and PLA-SP nanocomposites, an exothermic cold crystallization peak was observed at 100–121 and 105–125 °C, respectively. Moreover, an endothermic peak (relating melting point) was observed at about 148–151 °C for both neat and biocomposite films. In addition, the effectiveness of SP on cold crystallization kinetics of PLA was studied by utilizing Avrami model. The obtained Avrami coefficients for the modeling were found to be in the range of ~2.5–4. Moreover, Tobin model was used to interpret the nucleation growth mechanism in nonisothermal crystallization of neat PLA and prepared nanocomposite film. Further, the activation energy of crystallization was determined by employing Kissinger model and the nucleation growth was observed using POM. It is noteworthy to mention that even 5% loading of SP was capable to reduce OTR by more than 60%. Considering the above-discussed study, SP-incorporated PLA film can be used in the field of food packaging with better properties.

4.7 Summary

In this way, PLA with other renewable and biodegradable nanofillers is widely used for substituting various fossil-based plastics. Moreover, due to its tuneable biodegradability, nontoxicity and biocompatibility, PLA is widely used in various biomedical-related areas such as drug delivery, orthopedic fixation devices and tissue engineering as scaffolds. PLA and its copolymers are also being used to prepare a range of biomedical devices such as surgical sutures and healing dental wounds. Research area in PLA is now more focused for being a potential candidate in food-packaging industries in the form of biodegradable and compostable disposable cups for cold beverages, mugs, trays, plates and so on. Its unique physical properties also make it useful in diverse applications, including paper coating, fibers and films. The PLA can also be used in textile industries as woven and nonwoven fabrics, and further utilized in 3D printing also.

References

[1] Babu, RP., O'connor, K., & Seeram, R. Current progress on bio-based polymers and their future trends. Prog Biomater 2013, 2, 8.
[2] Garlotta, D. A literature review of poly (lactic acid). J Polym Environ 2001, 9, 63–84.
[3] Conn, RE., Kolstad, JJ., Borzelleca, JF., Dixler, DS., Filer Jr, LJ., LaDu Jr, BN., & Pariza, MW. Safety assessment of polylactide (PLA) for use as a food-contact polymer. Food Chem Toxicol 1995, 33, 273–283.
[4] Martinez, FAC., Balciunas, EM., Salgado, JM., González, JMD., & Converti, A. de Souza Oliveira RP. Lactic acid properties, applications and production: A review. Trends Food Sci Technol 2013, 30, 70–83.
[5] Yoo, DK., Kim, D., & Lee, DS. Synthesis of lactide from oligomeric PLA: Effects of temperature, pressure, and catalyst. Macromol Res 2006, 14, 510–516.
[6] Lunt, J. Large-scale production, properties and commercial applications of polylactic acid polymers. Polym Degrad Stab 1998, 59, 145–152.
[7] Gupta, AP., & Kumar, V. New emerging trends in synthetic biodegradable polymers–Polylactide: A critique. Eur Polym J 2007, 43, 4053–4074.
[8] Enomoto, K., Ajioka, M., & Yamaguchi, A. U.S. Patent No. 5,310,865. Washington, DC: U.S. Patent and Trademark Office1994.
[9] Myers, M., Connor, EF., Glauser, T., Möck, A., Nyce, G., & Hedrick, JL. Phosphines: Nucleophilic organic catalysts for the controlled ring-opening polymerization of lactides. J Polym Sci A 2002, 40, 844–851.
[10] Cheshmedzhieva, D., Angelova, I., Ilieva, S., Georgiev, GS., & Galabov, B. Initiation of ring-opening polymerization of lactide: The effect of metal alkoxide catalyst. Comput Theor Chem 2012, 995, 8–16.
[11] Dijkstra, PJ., Du, H., & Feijen, J. Single site catalysts for stereoselective ring-opening polymerization of lactides. Polym Chem 2011, 2, 520–527.
[12] Lee, HW., Insyani, R., Prasetyo, D., Prajitno, H., & Sitompul, JP. Molecular weight and structural properties of biodegradable PLA synthesized with different catalysts by direct melt polycondensation. J Eng Technol Sci 2015, 47, 364–373.
[13] Xiao, G., Yan, B., Ma, R., Jin, WJ., Lü, XQ., Ding, LQ., & Bao, F. Bulk ring-opening polymerization (ROP) of L-lactide catalyzed by Ni (II) and Ni (II)-Sm (III) complexes based on a salen-type Schiff-base ligand. Polym Chem 2011, 2, 659–664.

[14] Ray, L., Katiyar, V., Raihan, MJ., Nanavati, H., Shaikh, MM., & Ghosh, P. First example of a gold (I) N-heterocyclic-carbene-based initiator for the bulk ring-opening polymerization of l-lactide. Eur J Inorg Chem 2006, 692, 3724–3730.
[15] Samantaray, MK., Katiyar, V., Pang, K., Nanavati, H., & Ghosh, P. Silver N-heterocyclic carbene complexes as initiators for bulk ring-opening polymerization (ROP) of l-lactides. J Organomet Chem 2007, 692, 1672–1682.
[16] Marques, DA., Jarmelo, S., Baptista CM, G., & Gil, MH. Poly (lactic acid) synthesis in solution polymerization, Macromolecular symposia, Weinheim, WILEY-VCH Verlag, 2010, Vol. 296, 63–71.
[17] Ahmed, J., & Varshney, SK. Polylactides – Chemistry, properties and green packaging technology: A review. Int J Food Prop 2011, 14, 37–58.
[18] Auras, RA., Singh, SP., & Singh, JJ. Evaluation of oriented poly (lactide) polymers vs. existing PET and oriented PS for fresh food service containers. Packag Technol Sci 2005, 18, 207–216.
[19] Hamad, K., Kaseem, M., Yang, HW., Deri, F., & Ko, YG. Properties and medical applications of polylactic acid: A review. Express Polym Lett 2015, 9.
[20] Dhar, P., Kumar, A., & Katiyar, V. Magnetic cellulose nanocrystal based anisotropic polylactic acid nanocomposite films: Influence on electrical, magnetic, thermal, and mechanical properties. ACS Appl Mater Interfaces 2016, 8, 18393–18409.
[21] Dhar, P., Bhardwaj, U., Kumar, A., & Katiyar, V. Cellulose nanocrystals: A potential nanofiller for food packaging applications. Food Addit Packag 2014, 1162, 197–239.
[22] Dhar, P., Tarafder, D., Kumar, A., & Katiyar, V. Effect of cellulose nanocrystal polymorphs on mechanical, barrier and thermal properties of poly (lactic acid) based bionanocomposites. RSC Adv 2015, 5, 60426–60440.
[23] Ikada, Y., Jamshidi, K., Tsuji, H., & Hyon, SH. Stereocomplex formation between enantiomeric poly(lactides). Macromolecules 1987, 20, 904–906.
[24] Slager, J., & Domb, AJ. Biopolymer stereocomplexes. Adv Drug Deliv Rev 2003, 55, 549–583.
[25] Tsuji, H. Poly (lactide) stereocomplexes: Formation, structure, properties, degradation, and applications. Macromol Biosci 2005, 5, 569–597.
[26] Yui, N., Dijkstra, PJ., & Feijen, J. Stereo block copolymers of L-and D-lactides. Die Makromolekulare Chemie: Macromol Chem Phys 1990, 191, 481–488.
[27] Li, SH., & Woo, EM. Kinetic analysis on effect of poly (4-vinyl phenol) on complex-forming blends of poly (L-lactide) and poly (D-lactide). Polym J 2009, 41, 374.
[28] Chang, L., & Woo, EM. Effects of molten poly (3-hydroxybutyrate) on crystalline morphology in stereocomplex of poly (L-lactic acid) with poly (D-lactic acid). Polymer 2011, 52, 68–76.
[29] Sun, Y., & He, C. Synthesis and stereocomplex crystallization of poly (lactide)–Graphene oxide nanocomposites. ACS Macro Lett 2012, 1, 709–713.
[30] Quan, H., Zhang, SJ., Qiao, JL., & Zhang, LY. The electrical properties and crystallization of stereocomplex poly (lactic acid) filled with carbon nanotubes. Polymer 2012, 53, 4547–4552.
[31] Furuhashi, Y., Morioka, K., Tamegai, H., & Yoshie, N. Preparation and some properties of stereocomplex-type poly (lactic acid)/layered silicate nanocomposites. J Appl Polym Sci 2013, 127, 1615–1622.
[32] Bao, RY., Yang, W., Wei, XF., Xie, BH., & Yang, MB. Enhanced formation of stereocomplex crystallites of high molecular weight poly (l-lactide)/poly (d-lactide) blends from melt by using poly (ethylene glycol). ACS Sustain Chem Eng 2014, 2, 2301–2309.
[33] Purnama, P., & Kim, SH. Bio-based composite of stereocomplex polylactide and cellulose nanowhiskers. Polym Degrad Stab 2014, 109, 430–435.
[34] Han, L., Pan, P., Shan, G., & Bao, Y. Stereocomplex crystallization of high-molecular-weight poly (l-lactic acid)/poly (d-lactic acid) racemic blends promoted by a selective nucleator. Polymer 2015, 63, 144–153.

[35] Bor, Y., Alin, J., & Hakkarainen, M. Polylactide stereocomplexation leads to reduced migration during microwave heating in contact with food simulants. J Food Eng 2014, 134, 1–4.
[36] Bor, Y., Alin, J., & Hakkarainen, M. Electrospray ionization-mass spectrometry analysis reveals migration of cyclic lactide oligomers from polylactide packaging in contact with ethanolic food simulant. Packag Technol Sci 2012, 25, 427–433.
[37] Pal, AK., & Katiyar, V. Nanoamphiphilic chitosan dispersed poly (lactic acid) bionanocomposite films with improved thermal, mechanical, and gas barrier properties. Biomacromolecules 2016, 17, 2603–2618.
[38] Tripathi, N., & Katiyar, V. PLA/functionalized-gum arabic based bionanocomposite films for high gas barrier applications. J Appl Polym Sci 2016, 133.
[39] Valapa, R., Hussain, S., Iyer, PK., Pugazhenthi, G., & Katiyar, V. Non-isothermal crystallization kinetics of sucrose palmitate reinforced poly (lactic acid) bionanocomposites. Polym Bull 2016, 73, 21–38.

Prodyut Dhar and Vimal Katiyar

5 Polyhydroxyalkanoates: microbially derived biodegradable polymer for food packaging applications

5.1 Introduction

Polyhydroxyalkanoates (PHAs) are one of the classes of biodegradable polyesters produced through microbial fermentation routes utilizing renewable biomass resources, where the specific raw materials are processed into fermentable broths through chemical treatments. The production process of PHAs is usually accompanied by three steps: (i) preprocessing of the biomass into reduced sugars, which are easily fermentable as nutrient broth, (ii) optimization of the fermentation process for accumulation of PHAs under nutrient-stressed conditions and (iii) extraction and purification of the PHAs through cell lysis. PHAs are usually high-molecular-weight water-insoluble polymeric fragments present in the cytoplasm of microorganisms as intracellular granules of sizes in the range of 0.2–0.5 µm, and are synthesized by the genetically engineered or cloned bacterial strains [1]. Moreover, PHAs produced consist of various fractions and loading percentages of about 11 types of hydroxyalkanoate-based monomeric units, with some of the major prevailing fractions, namely 3-hydroxybutyrate (3-HB), 3-hydroxyvalerate (3-HV), 3-hydroxyhexanoate ($3HH_x$), 3-hydroxyoctanoate and 4-hydroxybutyrate (4-HB) with *R*-configuration of the stereospecificity [2]. These heterogeneous units undergo polymerization to form homopolymers as well as co- and heteropolymers with different compositions of molecular weights in the range of 200,000–3,000,000 Da [3]. Moreover, it was found that the fractions and type of PHAs produced apparently depend on the synthesis route, bacterial strain used for fermentation and physiological stresses introduced by varying the carbon and nitrogen ratio during growth conditions, thereby altering the physicochemical properties of the synthesized polymer [4]. In the early nineteenth century, presence of the sudanophilic, lipid-like PHAs were first observed in *Azotobacter chroococcum*, which was chemically thought to be building blocks of the 3-HB units especially accumulated by the Gram-negative bacteria of genus *Bacillus* [5]. For last few decades, extensive research has been carried out in engineering the microbial strains, fermenter studies and optimization of the purification process, which has successfully led to its feasibility for industrial–scale productions by renowned companies such as Mitsubishi Gas Chemical Company Inc, BASF (Germany), P&G (USA) and Jiangmen Biotech Ctr (China) with the production capacity of 50–50,000 tons/annum and commercialized under various trade names [6].

https://doi.org/10.1515/9783110648034-005

5.2 Polyhydroxyalkanoates: different types and their physicochemical properties

PHAs have major fraction of the hydroxyalkanoates units as 3-HB, and exist in homopolymeric units to form a brittle polymer, poly(3-hydroxybutyrate) (P(3HB) or PHB) which was first isolated and characterized by Maurice Lemoigne in the year 1925 [6]. PHBs, the most popular type of PHAs, produced by more than 75 genera of bacteria are isotactic in structure with very high crystallinity of 55–85%. PHBs typically have molecular weight in the range of 10–3,000 kDa with high melting temperature of (T_m) ~ 175–180 °C, glass transition temperature (T_g) ~4 °C and crystallization temperature (T_{cc}) ~102–106 °C, respectively. In view of mechanical properties, PHBs have higher tensile strength of ~43 MPa with the Young's modulus ~3.5 GPa, but are highly brittle with reduced elongation at break of ~5% due to their crystalline nature, which limits their applications [7]. Compared to other biodegradable polymers, PHBs are readily degradable with improved crystallization and thermal properties, but structural and barrier properties need to be improved through introduction of nanoparticles or biobased nanofillers to have a competitive market with conventional petroleum-derived polymers [8]. As discussed earlier, the monomeric units in PHB can be incorporated with other hydroxyalkanoate units to form a random copolymer of various chain lengths, which can be fabricated through tuned microbial fermentation or chemosynthesis process. Poly(3-hydroxybutyrate-co-3-hydroxyvalerate) (PHBV) is such a type of random copolymer made from 3-HB and 3-HV units, which shows isodimorphism characteristics; hence, it undergoes cocrystallization depending upon the relative content of hydroxyalkanoate units [9]. Such copolymeric units with variation in HV content by ~ 0–95 mol% are obtained by utilizing carbon sources such as propionic acid, pentanoic acid and butyric acids as starting materials. An increase in the 3-HV units reduces the crystallinity and rate of crystallization of PHAs; however, the biodegradability improves with enhancement in fractions of 3-HB [10]. By varying the % HV content, it is possible to tune the intrinsic properties of the PHBV such as glass transition temperature, melting temperature (130–170 °C), tensile strength (25–40 MPa) and elongation behavior (~20%), which is slightly improved in comparison to PHB [11]. Poly(3-hydroxybutyrate-co-4-hydroxybutyrate) (P3HB4HB) is another copolymer synthesized using monomers of 3-HB and 4-HB, respectively, which is produced when 1,4-butanediaol or 4-hydroxybutyric acid are used as starting materials for microbial fermentations and usually exists in single crystalline form of P(3HB) when 4-HB content is between ~0 and 29 mol% and with P(4HB) crystals when it has 4-HB content of ~78–100 mol% [12]. Increase in the % 4-HB content in the copolymer invariably reduces the rate of crystallization and crystallinity from 60% to 14%, improving the structural properties, however, decreasing the biodegradation rate [13]. Interestingly, variation of 3-HB and 4-HB units in the P3HB4HB copolymers leads to wide alteration in its properties with melting and glass transition temperatures decreasing from 178 to 130 °C and 4 to −48 °C, respectively, and tensile modulus in the range of 20–25 MPa, however, with significantly high elongation at break of 444% [14]. Introduction of $3HH_x$ units with propyl side groups leads to formation of the poly(3-hydroxybutyrate-co-3-hydroxyhexanoate) ($PHBHH_x$) copolyesters, which usually have 0–35 mol% of HH_x units in them. These copolyesters generally have tunable properties with decreased melting temperature of ~150 °C at 1.5 mol% of HH_x and 52 °C at 25 mol% of HH_x, glass transition temperature of −2 and −4 °C at 17 and 25 mol% HH_x loadings, respectively [15]. Compared to PHB and its derivatives, the processed films of $PHBHH_x$ are flexible with improved elongation at break of ~850% but with reduced tensile modulus of ~20 MPa [16]. The other PHA derivatives such as poly(3-hydroxyoctanoateco-3-hydroxyhexanoate) and poly(3-hydroxyoctanoateco-3-hydroxypropanoate) are some of the copolyesters that can be produced using the octanoic acid and 3-hydroxypropionoc acid, respectively, as carbon sources [17]. Therefore, from the above discussion it is conclusive that physicochemical properties of PHAs can be remotely regulated by controlling the concentrations of hydroxyalkanoate monomeric units during the microbial fermentation process, for the development of the value-added products based on the targeted applications (as shown in Figure 5.1).

PHA-based derivatives produced through tuned microbial fermentations have been widely used for fabrication of nanocomposites for potential range of applications such as biodegradable food packaging, in tissue engineering as scaffolds due to their improved biocompatibility or as targeted drug delivery-based biomedical devices [18]. In this section, we present an overview of recent advances in the field of PHA-based nanocomposite development and their evaluation of properties for overcoming the associated technoeconomical shortcomings. Two classes of PHA-based polymers such as P(3HB) and PHBV have been widely studied in literature for the fabrication of the nanocomposites through several processing technologies such as solution-casting and extrusion-based molding technologies. Processing of PHAs through industrially viable extrusion is a challenging task due to their reduction in molecular weight under high shear and temperature conditions, which deteriorates the inherent characteristics of the polymer required for the development of value-added products [19]. Moreover, the PHAs generally have low structural and gas barrier properties required for packaging applications along with poor film-forming ability due to wide variation in their molecular weight as well as hydroxyalkanoate compositions. To overcome such deficiencies, micro/nanofillers dispersed into the polymeric matrix at percolated threshold concentrations have been instrumental in improving the properties of polymeric nanocomposites. Inorganic nanofillers such as montmorillonite and its chemically modified forms such as neopentyl(diallyl)oxy tri (dioctyl)pyrophosphate titanate have been shown to improve the dispersion of nanoclays and act as a strong reinforcing agents [20]. PHB/modified montmorillonite nanoclay-based composites prepared through melt extrusion leads to intercalation of polymer into clay galleries, which enhances the Young's modulus and lowers the biodegradation rate [21]. Other clay derivatives such as kaolinite under melt intercalation

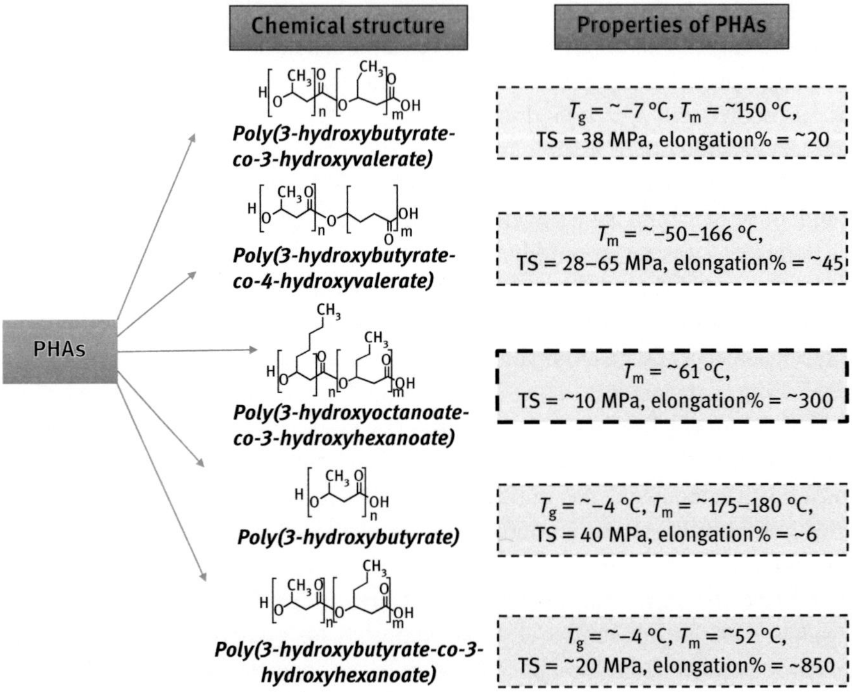

Figure 5.1: The different fractions of the PHAs produced through microbial fermentation route with their respective physicochemical, structural and thermal properties.

lead to expansion of the polymer into clay galleries, which significantly enhance the thermal stability of the prepared nanocomposites [22]. A recent study by Hema et al. [23] utilized different forms of PHAs such as P(3HB), poly(3-hydroxybutyrate-co-6%3-hydroxyvalerate) copolymer, poly(3-hydroxybutyrate-co-70%4-hydroxybutyrate) copolymer and poly(3-hydroxybutyrate-co-10%3-hydroxyvalerate-co-10%4-hydroxybutyrate) terpolymer for the preparation of nanocomposite films using the commercial clays such as Closite 30B and Claytone APA. The solution-processed films show enhanced thermomechanical properties along with antimicrobial activity, making it ideal for packaging applications especially for storage of food products. Blending of the noncompatible polymers such as poly(butylene adipate-co-terephthalate) through the incorporation of the functionalized montmorillonite acts as the internetwork with the PHBV polymeric matrix, improving their miscibility as well as crystallization and thermal properties [24]. Sepiolite-based nanoclays with the silanol functional groups improve the dispersibility, reducing the thermally induced molecular weight degradation, which leads to improvement in mechanical and thermal resistance of the P(3HB-co-4HB)-based nanocomposites [25]. Incorporation of the silane or amine-modified halloysite nanoclays have been shown to have interesting interfacial interaction with the carboxyl groups present in PHBV, which are found to be dependent on the location of functional moieties in nanofillers [26]. Utilization of the unpurified PHBV derived from microbial fermentation directly shows improved water vapor permeability with high flexibility which reduces the cost of purification of the PHAs as well as provides an affordable method to derive packaging products from wastes [27]. The biobased foams derived using PHBV/clay nanocomposites through continuous supercritical CO_2-assisted extrusion process leads to the formation of biodegradable foams with high porosity and thermal stability required for engineering applications as a replacement to the synthetic foams derived from fossil fuels [28]. PHAs are known for their easy biodegradability that are known to vary depending on the presence of copolymers of HAs, which were recently studied by Ong and Sudesh, utilizing four types of PHAs such as P(3HB), poly(3HB-co-8 mol% 3-hydroxyhexanoate) [P(3HB-co-8 mol% $3HH_x$)], P(3HB-co-12 mol% $3HH_x$) and P(3HB-co-21 mol% $3HH_x$) [29]. All the PHA derivative films when kept in the soil condition degraded within 8 weeks, which showed drastic reduction in number and average molecular weight, with the highest degradation rate for the copolymer P(3HB-co-21 mol% $3HH_x$). After the degradation of PHA films, the microbial community in the soil was found to be significantly altered, which might be helpful in improving soil fertility and requires further characterizations [29].

PHA-based biodegradable films with the incorporation of advanced functional nanomaterials such as metallic or bioderived nanoparticles lead to integration of smart-responsive characteristics along with a strong reinforcement effect for potential application of PHA films as active packaging. Biodegradable PHA films derived from renewable resources incorporated with antimicrobial/antibacterial or biocide effect for packaging purposes are of keen interest, which will increase the shelf life of stored food items. Silver nanoparticles incorporated poly(3-hydroxybutyrate-co-18 mol%-3-hydroxyvalerate) (PHBV18) films at very low loading fractions show significantly improved barrier properties along with prolonged antimicrobial performance for almost 7 months against two pathogens, *Salmonella enterica* and *Listeria monocytogenes*, found commonly in foods as contaminants [30]. In another study, antimicrobial activity was incorporated into PHB films through Ag_2S particles, which were transferred to the organic polymer solution from aqueous phase through a two-step process and are also found to increase the thermal degradation rate [31]. Micron- and nanosized zinc oxide particles of various morphologies were evaluated for their performance as bactericide when incorporated into PHBV through different methods such as direct melt extrusion or preincorporation of the zinc oxide into PHBV18 through electrospinning or as coating of PHBV18/zinc oxide particle electrospun mats onto PHBV films through compression molding. The antibacterial activity was found to be based on the specific surface area as well as morphology type, with the hexagonal-pyramid zinc oxide nanoparticles showing the highest biocide behavior against the well-known food pathogen *Listeria monocytogenes* [32]. The different methodologies used to incorporate zinc oxide crystals into polymeric films were to highlight the potential end-use applications, for example, the packaging for storage of fruits and vegetables should have enough oxygen permeation to keep products fresh and for modified atmosphere packaging should have high oxygen barrier properties. It was found that for almost all the cases mentioned earlier, the incorporated inorganic metal oxide nanoparticles show their antibacterial activity for storage of food products with variation in its aeration conditions. An interesting study reported by Elias and coworkers [33] utilized the bacteriophages (viruses that attack bacteria) to specifically detect the bacterial moieties that were growing as contaminants on the biopolymers followed by its recognition and disinfection. Conjugation of phases to the surface of the PHA films was carried out through simple oxygen-plasma treatment followed by their exposure to host bacteria *Escherichia coli*, which showed two orders of the bacterial growth inhibition.

The carbon-based nanomaterials such as graphene, graphene oxide, carbon nanotubes and its allotropes are known to enhance the conductivity behavior in the incorporated nanocomposites due to which they have diverse applications as sensor-based packaging devices. An ecofriendly photoresponsive device was prepared utilizing transparent PHB/carbon nanotube films, which were deposited with the conducting polymers poly-*N*–90-heptadecanyl-2,7-carbazole-alt-5,5-(40,70-di-2-thienyl-20,10,30-benzothiadiazole) that acted as an electron donor and an electron acceptor, that is, phenyl-C70-butyric acid methyl ester (PC70BM) [34]. The multiwalled carbon nanotubes with various aspect ratio and loading concentrations when introduced into the poly(hydroxyl butyrate-co-valerate) (PHBV) and poly(butylenes adipate-co-terephthalate) (PBAT) blends showed significant variation in the electrical resistivity as well as dielectric responses, which could be used for development of stress-controlled sensors of high energy storage materials [35]. A recent study showed that spindle-like hierarchical carbon structure can be grown through thermal decomposition of the PHA/ferrocene/chloroform precursor, which showed high capacitance with excellent recycling stability. The PHA/carbon nanotube composites when seeded with the cells followed by its electric stimulation also showed the ability to promote differential cell viability that mades it suitable for development of neural interface biomaterials [36]. Moreover, along with the introduction of sensing capability, the carbon nanotubes also showed significant improvement in the structural and thermal stability of the processed PHB nanocomposites [37]. Manufacturing of such stimuli-responsive biodegradable and ecofriendly films with the ability to detect the impurities, contaminants and monitoring the extent of deterioration in food quality with time during the storage is required for development of smart packaging devices.

Fabrication of the food packages with the inorganic metallic nanofillers and carbon-based nanomaterials (even at low loadings) is known to have concerns regarding their degree of toxicity, problems related to biodegradability, environmental risk factors and its biocompatibility issues (at various concentration ranges), which are still not well understood. Henceforth, alternative strategies to develop completely green and sustainable packaging using biopolymers or bionanofillers derived from renewable resources as reinforcing agents can effectively overcome the associated problems and concerns (as mentioned earlier). Biopolymers such as cellulose, starch, chitin, chitosan and keratin, which are abundantly available in nature in its pure form or as derivatives, can be effectively extracted through pretreatment process followed by its dispersion into PHAs directly or through strategic surface modification approaches. Chitosan, the acetylated form of chitin (generally derived from the sea weeds or crab shells), is cross-linked with PHBV to form scaffolds or electrospun nanofibrous mats, which shows improved biocompatibility with improved cell adhesion and proliferation for the repair of nerve defects [38, 39]. Along with the enhanced biocompatibility, dispersion of the chitosan or the antioxidant catechin leads to improved mechanical and thermal properties with the enhanced biodegradation behavior [40]. The chitosan-based micro/nanoparticles are also loaded with antioxidants, drugs or other such biomolecules and their sustained release through PHA films surface might increase the shelf life of stored food items in contact with the surface of films [41]. Starch is generally amylose and amylopectin units with various compositions generally derived from green plants, when blended with biodegradable polymers such as PHB, which significantly alters its properties depending on the fractions of two units. Starch with high content of amylose units leads to improved hydrogen bonding with the polymeric matrix, thereby enhancing the thermal, rheological and mechanical properties of the PHB/starch composites [42]. PHBV blended with starch at high loading fractions (of ~50 wt%) becomes completely immiscible with lack of interfacial adhesion [43] but reactive compatibilization in the presence of maleic anhydride and benzoyl peroxide leads to formation of reduced particle size of starch along with its improved dispersion in PHB matrix [44]. Keratins are fibrous structural proteins, generally derived from hair, skins or nails of mammals through a stringent pre-treatment and purification process, which converts it to nanodimensions and after blending it with PHAs through melt extrusion or through electrospinning shows significantly high resistance to the permeation of oxygen [45].

5.3 Polyhydroxyalkanoate-based cellulosic composites

Cellulose is one of the most available biopolymers, which are extracted from renewable biomass resources through several pretreatment procedures and can be converted into forms of either micro- or nanomorphologies such as cellulose fibers, microcrystalline cellulose, nanocellulose, cellulose nanowhiskers and cellulose nanocrystals (CNCs). CNCs developed from the renewable biomass resources are smart bionanoparticles with attractive anisotropic mechanical properties (comparable to steel with $E_{||}$ = 110–220 GPa and $E_{\perp}$ = 2–50 GPa), high hydroxyl functionality, rod-like morphology with tunable aspect ratio and nontoxicity make it an ideal candidate for reinforcing agent in packaging materials. In one of our recent study,

CNCs extracted from crystalline domains of pretreated cellulose present in bamboo stems underwent improved dispersion in the PHB matrix when dispersed through solvent-*cum*-solution casting approach. Presence of high hydroxyl functionality in CNCs leads to the formation of the percolated network-like structure in polymer matrix through the formation of inter- or intramolecular hydrogen bonding. This leads to the improvement in the tensile strength and the Young's modulus by ~58% for PHB/CNC nanocomposites [46] and, on the other hand, incorporation of PEG as plasticizer at low fractions leads to increment in elongation behavior by approximately ~50 times [47]. CNCs derived from microcrystalline cellulose for preparation of the PHB/CNC nanocomposites have improved resistance to water vapor permeation along with the increased UV barrier properties more than that of the synthetic polymers such as propylene [48]. The solution-processed PHB/CNC films are highly transparent with improved physicochemical, thermal, barrier and structural properties and interestingly have overall migration values within the standard limits as per the set legislations required for stringent food packaging [49]. Moreover, the detailed linear and nonlinear rheological investigations helped to understand the stability of percolated CNC/PHB network formed under small and large amplitude oscillatory stress conditions [6]. The stability of network was projected through the permeation activation energy values and the permeability parameters such as solubility and diffusivity coefficients, respectively, which were measured to physically understand the effect of tortuosity or physicochemical interaction between PHB/CNC that were responsible for reduced oxygen transport across the films. The reduction in the oxygen transmission rate was observed to be ~65%, which was predominantly governed due to reduction in solubility due to formation of the percolated structure through hydrogen bonding between the CNC and PHB matrix (as shown in Figure 5.2), which restricted the permeation of oxygen gas molecules across film [49]. Moreover, electrospun fiber of the PHB/CNCs showed improved biocompatibility toward L929 cell line with improved adhesion and cell proliferation, which suggested the nontoxic characteristics of the PHB/CNC films [50]. Therefore, it could be concluded that incorporation of CNCs at very low fractions (~1–2 wt.%) significantly improves the inherent properties of PHB films along with the integration of complete biodegradability, nontoxicity and biocompatibility, which makes it a potential alternative to the nondegradable petroleum-derived packages.

PHB/CNC nanocomposites have been blended with other biodegradable polymers such as PLA to form ternary composites, which are extensively studied in the literature due to their improved compatibilization when processed through different techniques. CNCs are known to act as a reinforcing agent between the PLA and PHB matrix, thereby increasing the interfacial adhesion between the two polymeric matrixes at a ratio of 70/30, which thereby improves the flexural modulus by ~104% and reduces oxygen permeability by ~27%, respectively [51]. Whereas at PLA/PHB ratio of 71.25/23.75 wt% with 5 wt% of incorporated CNCs, the elongation at break significantly increases to ~175%, with enhanced disintegration rate of ~10 days

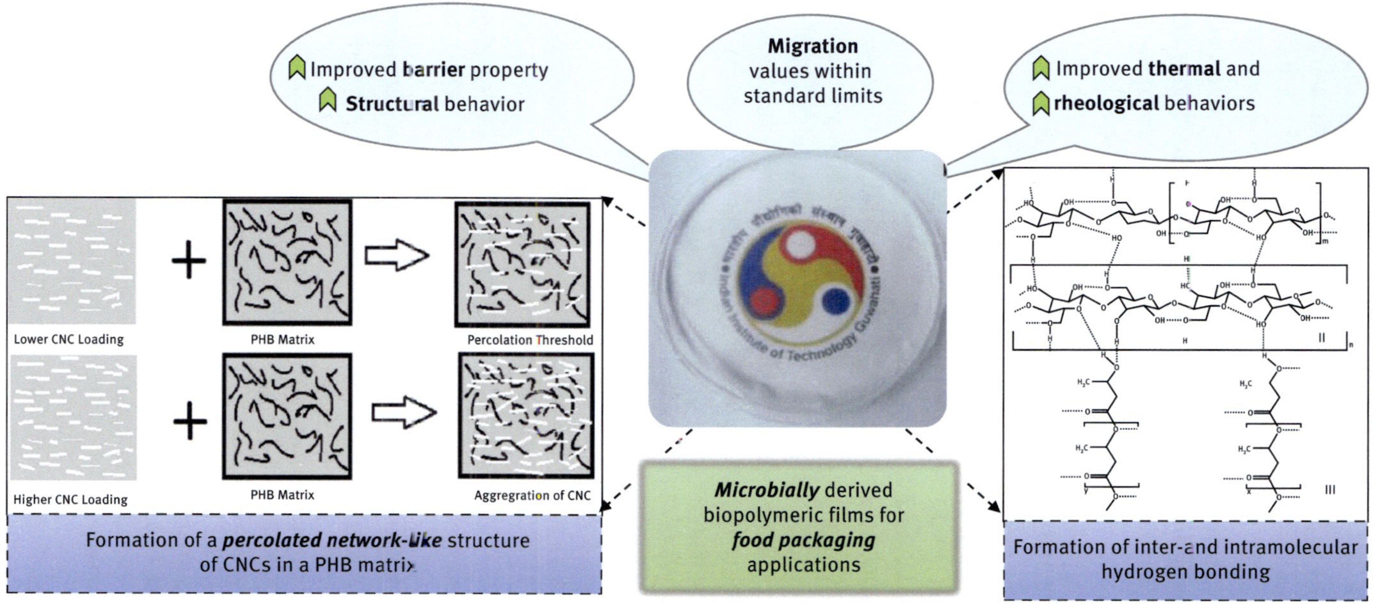

Figure 5.2: Formation of the percolated-network-like structure through the presence of inter- or intramolecular hydrogen bonding between the CNCs and PHB matrix, which leads to significant improvement in the structural properties, thermal behavior, reduced permeation of oxygen gas and is within standard migration limits required for stringent food packaging applications.

under composting conditions [52]. Introduction of the high aspect ratio cellulose nanofibers into the PLA matrix with PHB as carrier systems shows the formation of shear-thinning behavior as confirmed from rheological studies with significant improvement in the tensile and flexural modulus [53]. Electrospinning of PLA/PHB nanomats at a ratio of 60:20, with the addition of 5 wt% CNC and 15 wt% acetyl (tributyl citrate) as plasticizer leads to the formation of uniform morphology and bead-free fibers with higher flexibility and thermal properties along with enhanced composting rate of ~10 days [54]. In another study, Kenny and coworkers [55] show that melt blending of the PLA/PHB blends at a ratio of 71.25/23.75 wt% with the addition of 5 wt% CNCs forms a bridge between the two polymeric matrix to improve their interfacial adhesion. The fabricated ternary nanocomposites show improved processability and subsequently incorporate synergistic improvement in the crystallinity and thermal properties. As discussed in earlier section, PHAs tend to have poor film-forming ability and extrudability due to enhanced degradation in molecular weight during processing, which makes it very difficult to fabricated PHA-based nanocomposites for packaging applications at commercial scale. Formation of the ternary blends with PLA and CNCs not only improves the processability but also provides structural stability, reducing the oxygen and water vapor permeabilities and most importantly retaining the biodegradability. Therefore, it could be observed from above discussion that CNCs not only act as a strong reinforcing agent but also form a compatibilizer between the two polymeric systems in the fabricated ternary nanocomposites whose properties can be finely tuned through alteration in PLA/PHB ratio depending upon the targeted end-user applications.

References

[1] Dawes, EA., & Senior, PJ. The role and regulation of energy reserve polymers in microorganisms polyhydroxybutyrate. Adv Microb Physiol 1973, 10, 203–203.

[2] Reddy, CSK., Ghai, R., & Kalia, V. Polyhydroxyalkanoates: an overview. Bioresour Technol 2003, 87, 137–146.

[3] Sudesh, K., Abe, H., & Doi, Y. Synthesis, structure and properties of polyhydroxyalkanoates: biological polyesters. Prog Polym Sci 2000, 25, 1503–1555.

[4] Scaffaro, R., Dintcheva, NT., Marino, R., & Mantia, FPL. Processing and properties of biopolymer/polyhydroxyalkanoates blends. J Polym Environ 2012, 20, 267–272.

[5] Meyer, A. Practicum der botanischen bakterienkunde. Jena: Gustav-Fischer 1903, 2.

[6] Bhardwaj, U., Dhar, P., Kumar, A., & Katiyar, V. Polyhydroxyalkanoates (PHA)-Cellulose Based Nanobiocomposites for Food Packaging Applications, In: Food Additives and Packaging, Eds, Vanee Komolprasert, Petra Turowski, ACS Symp Ser, 2014, Vol. 1162, 275–314.

[7] Suzuki, T., Yamane, T., & Shimizu, S. Mass production of poly-β-hydroxybutyric acid by fed-batch culture with controlled carbon/nitrogen feeding. Appl Microbiol Biotechnol 1986, 24, 370–374.

[8] Hartmann, R., Hany, R., Geiger, T., Egli, T., Witholt, B., & Zinn, M. Tailored biosynthesis of olefinic medium-chain-length poly [(R)-3-hydroxyalkanoates] in pseudomonas putida GPo1 with improved thermal properties. Macromolecules 2004, 37, 6780–6785.
[9] Bloembergen, S., Holden, DA., & Hamer, GK. Bluhm TL and Marchessault RH, Studies of composition and crystallinity of bacterial poly (β-hydroxybutyrate-co-β-hydroxyvalerate). Macromolecules 1986, 19, 2865–2871.
[10] VanderHart, DL., Orts, WJ., & Marchessault, RH. 13C NMR determination of the degree of cocrystallization in random copolymers of poly (beta.-hydroxybutyrate-co-. beta.-hydroxyvalerate). Macromolecules 1995, 28, 6394–6400.
[11] Barham, PJ., Barker, P., & Organ, SJ. Physical properties of poly (hydroxybutyrate) and copolymers of hydroxybutyrate and hydroxyvalerate. FEMS Microbiol Lett 1992, 9, 289–298.
[12] Kunioka, M., Kawaguchi, Y., & Doi, Y. Production of biodegradable copolyesters of 3-hydroxybutyrate and 4-hydroxybutyrate by Alcaligenes eutrophus. Appl Microbiol Biotechnol 1989, 30, 569–573.
[13] Kimura, H., Yoshida, Y., & Doi, Y. Production of poly (3-hydroxybutyrate-co-4-hydroxybutyrate) by Pseudomonas acidovorans. Biotechnol Lett 1992, 14.
[14] Saito, Y. Nakamura S and Hiramitsu M, Microbial synthesis and properties of poly (3-hydroxybutyrate-co-4-hydroxybutyrate). Polym Int 1996, 39, 169–174.
[15] Doi, Y. Kitamura S and Abe H, Microbial synthesis and characterization of poly (3-hydroxybutyrate-co-3-hydroxyhexanoate). Macromolecules 1995, 28, 4822–4828.
[16] Fukui, T., Abe, H., & Doi, Y. Engineering of Ralstonia eutropha for production of poly (3-hydroxybutyrate-co-3-hydroxyhexanoate) from fructose and solid-state properties of the copolymer. Biomacromolecules 2002, 3, 618–624.
[17] Ihssen, J., Magnani, D., Thöny-Meyer, L., & Ren, Q. Use of extracellular medium chain length polyhydroxyalkanoate depolymerase for targeted binding of proteins to artificial poly [(3-hydroxyoctanoate)-co-(3-hydroxyhexanoate)] granules. Biomacromolecules 2009, 10, 1854–1864.
[18] Keshavarz, Philip, S., & Roy, I. Polyhydroxyalkanoates: Biodegradable polymers with a range of applications. J Chem Technol Biotechnol 2007, 82, 233–247.
[19] Ray, SS., & Bousmina, M. Biodegradable polymers and their layered silicate nanocomposites: In greening the twenty-first century materials world. Prog Mater Sci 2005, 50, 962–1079.
[20] Parulekar, Y. Mohanty AK and Imam SH, Biodegradable nanocomposites from toughened polyhydroxybutyrate and titanate-modified montmorillonite clay. J Nanosci Nanotechnol 2007, 7, 3580–3589.
[21] Botana, A., & Mollo, M. Eisenberg P and Torres Sanchez RM, Effect of modified montmorillonite on biodegradable PHB nanocomposites. Appl Clay Sci 2010, 47, 263–270.
[22] Gardolinski, JE., Carrera, LCM., Cantão, MP., & Wypych, F. Layered polymer-kaolinite nanocomposites. J Mater Sci 2000, 35, 3113–3119.
[23] Hema, R., Ng, PN., & Amirul, AA. Green nanobiocomposite: Reinforcement effect of montmorillonite clays on physical and biological advancement of various polyhydroxyalkanoates. Polymer Bull 2013, 70, 755–771.
[24] Bittmann, B., Bouza, R., Barral, L., Castro-Lopez, M., & Dopico-Garcia, S. Morphology and thermal behavior of poly (3-hydroxybutyrate-co-3-hydroxyvalerate)/poly (butylene adipate-co-terephthalate)/clay nanocomposites. Composites 2015, 36, 2051–2058.
[25] Khandal, D., Pollet, E., & Avérous, L. Elaboration and behavior of poly (3-hydroxybutyrate-co-4-hydroxybutyrate)-nano-biocomposites based on montmorillonite or sepiolite nanoclays. Eur Polym J 2016, 81, 64–76.

[26] Daitx, TS., Carli, LN., Crespo, JS., & Mauler, RS. Effects of the organic modification of different clay minerals and their application in biodegradable polymer nanocomposites of PHBV. Appl Clay Sci 2015, 115, 157–164.
[27] Martínez-Abad, A., Cabedo, L., Oliveira, CSS., Hilliou, L., Reis, M., & Lagarón, JM. Characterization of polyhydroxyalkanoate blends incorporating unpurified biosustainably produced poly (3-hydroxybutyrate-co-3-hydroxyvalerate). J Appl Polym Sci 2016, 133.
[28] Le Moigne, N., Sauceau, M., Benyakhlef, M., Jemai, R., Benezet, J.C., Elisabeth, R., José-Marie, L.C., & Jacques, F. Foaming of poly(3-hydroxybutyrate-co-3-hydroxyvalerate)/organo-clays nano-biocomposites by a continuous supercritical CO2 assisted extrusion process. Eur Polym J 2014, 61, 157–171.
[29] Ong, SY., & Sudesh, K. Effects of polyhydroxyalkanoate degradation on soil microbial community. Polym Degrad Stab 2016, 131, 9–19.
[30] Castro-Mayorga, JL., Fabra, MJ., & Lagaron, JM. Stabilized nanosilver based antimicrobial poly (3-hydroxybutyrate-co-3-hydroxyvalerate) nanocomposites of interest in active food packaging. Innov Food Sci Emerg Technol 2016, 33, 524–533.
[31] Yeo, SY., Tan, WL., Abu Bakar, M., & Ismail, J. Silver sulfide/poly (3-hydroxybutyrate) nanocomposites: thermal stability and kinetic analysis of thermal degradation. Polym Degrad Stab 2010, 95, 1299–1304.
[32] Castro-Mayorga, JL., Fabra, MJ., Pourrahimi, AM., Olsson, RT., & Lagaron, JM. The impact of zinc oxide particle morphology as an antimicrobial and when incorporated in poly (3-hydroxybutyrate-co-3-hydroxyvalerate) films for food packaging and food contact surfaces applications. Food Bioprod Process 2017, 101, 32–44.
[33] Wang, C., Sauvageau, D., & Elias, A. Immobilization of active bacteriophages on polyhydroxyalkanoate surfaces. ACS Appl Mater Interfaces 2016, 8, 1128–1138.
[34] Valentini, L., Fabbri, P., Messori, M., Degli Esposti, M., & Bittolo Bon, S. Multilayer films composed of conductive poly (3-hydroxybutyrate)/carbon nanotubes bionanocomposites and a photoresponsive conducting polymer. J Polym Sci B 2014, 52, 596–602.
[35] Musto, P., Russo, P., Cimino, F., & Acierno, D. Lupò G and Petrarca C, Dielectric behavior of biopolymer based composites containing multi wall carbon nanotubes: effect of filler content and aspect ratio. Eur Polym J 2015, 64, 170.
[36] Vallejo-Giraldo, C., Pugliese, E., Larrañaga, A., Fernandez-Yague, MA., Britton, JJ., Trotier, A., Tadayyon, G., Kelly, A., Rago, I., Sarasua, JR., & Dowd, E. Polyhydroxyalkanoate/carbon nanotube nanocomposites: Flexible electrically conducting elastomers for neural applications. Nanomedicine 2016, 11, 2547–2563.
[37] Lemes, AP., Montanheiro, TLA., Passador, FR., & Durán, N. Nanocomposites of Polyhydroxyalkanoates Reinforced with Carbon Nanotubes: Chemical and Biological Properties, In: Eco-friendly Polymer Nanocomposites, editors, Thakur VK, Thakur MK,. Springer, India, 2015.
[38] Veleirinho, B., Coelho, DS., Dias, PF., Maraschin, M., Ribeiro-do-Valle, RM., & Lopes-da-, JA. Silva Nanofibrous poly (3-hydroxybutyrate-co-3-hydroxyvalerate)/chitosan scaffolds for skin regeneration. Int J Biol Macromol 2012, 51, 343–350.
[39] Keshel, SH., Biazar, E., Rezaei Tavirani, M., Rahmati Roodsari, M., Ronaghi, A., Ebrahimi, M., … & Afsordeh, K. The healing effect of unrestricted somatic stem cells loaded in collagen-modified nanofibrous PHBV scaffold on full-thickness skin defects. Artif cells, nanomed biotechnol 2014, 42, 210–216.
[40] Arrieta, MP., López, J., López, D., Kenny, JM., & Peponi, L. Effect of chitosan and catechin addition on the structural, thermal, mechanical and disintegration properties of plasticized electrospun PLA-PHB biocomposites. Polym Degrad Stab 2016, 132, 145–156.

[41] Lins, LC., Padoin, N., Pires, ATN., & Soares, C. Modeling ketoprofen release from PHB/chitosan composite microparticles. Polym Bull 2016, 73, 1515–1529.
[42] Zhang, M., & Thomas, NL. Preparation and properties of polyhydroxybutyrate blended with different types of starch. J Appl Polym Sci 2010, 116, 688–694.
[43] Reis, KC., Pereira, J., Smith, AC., Carvalho, CWP., Wellner, N., & Yakimets, I. Characterization of polyhydroxybutyrate-hydroxyvalerate (PHB-HV)/maize starch blend films. J Food Eng 2008, 89, 361–369.
[44] Ma, P., Xu, P., Chen, M., Dong, W., Cai, X., & Schmit, P. Structure–Property relationships of reactively compatibilized PHB/EVA/starch blends. Carbohydr Polym 2014, 108, 299–306.
[45] Fabra, MJ., Pardo, P., Martínez-Sanz, M., Lopez-Rubio, A., & Lagarón, JM. Combining polyhydroxyalkanoates with nanokeratin to develop novel biopackaging structures. J Appl Polym Sci 2016, 133.
[46] Dhar, P., Bhardwaj, U., Kumar, A., & Katiyar, V. Investigations on rheological and mechanical behavior of poly (3-Hydroxybutyrate)/cellulose nanocrystal based nanobiocomposites. Polym Compos 2017, 38, E392–E401.
[47] Patrício, PS., Pereira, FV., Dos Santos, MC., de Souza, PP., Roa, JPB., & Orefice, RL. Increasing the elongation at break of polyhydroxybutyrate biopolymer: Effect of cellulose nanowhiskers on mechanical and thermal properties. J Appl Polym Sci 2013, 127, 3613–3621.
[48] Seoane, IT., Fortunati, E., Puglia, D., Cyras, VP., & Manfredi, LB. Development and characterization of bionanocomposites based on poly (3-hydroxybutyrate) and cellulose nanocrystals for packaging applications. Polym Int 2016, 65, 1046–1053.
[49] Dhar, P., Bhardwaj, U., Kumar, A., & Katiyar, V. Poly (3-hydroxybutyrate)/cellulose nanocrystal films for food packaging applications: barrier and migration studies. Polym Eng Sci 2015, 55, 2388–2395.
[50] Kampeerapappun, P. The electrospun polyhydroxybutyrate fibers reinforced with cellulose nanocrystals: morphology and properties. J Appl Polym Sci 2016, 133.
[51] Dasan, YK., Bhat, AH., & Ahmad, F. Polymer blend of PLA/PHBV based bionanocomposites reinforced with nanocrystalline cellulose for potential application as packaging material. Carbohydr Polym 2017, 157, 1323–1332.
[52] Arrieta, MP., Fortunati, E., Dominici, F., Rayón, E., López, J., & Kenny, JM. PLA-PHB/cellulose based films: mechanical, barrier and disintegration properties. Polym Degrad Stab 2014, 107, 139–149.
[53] Kiziltas, A., Nazari, B., Erbas Kiziltas, E., Gardner, D., Han, Y., & Rushing, TS. Cellulose NANOFIBER-polyethylene nanocomposites modified by polyvinyl alcohol. Carbohydr Polym 2016, 133.
[54] Arrieta, MP., López, J., López, D., Kenny, JM., & Peponi, L. Biodegradable electrospun bionanocomposite fibers based on plasticized PLA–PHB blends reinforced with cellulose nanocrystals. Ind Crops Prod 2016, 93, 290–301.
[55] Arrieta, MP., Fortunati, E., Dominici, F., Rayón, E., López, J., & Kenny, JM. Multifunctional PLA–PHB/cellulose nanocrystal films: processing, structural and thermal properties. Carbohydr Polym 2014, 107, 16–24.

Surendra Singh Gaur, Tabli Ghosh and Vimal Katiyar

6 General material property requirement for food packaging applications

6.1 Introduction

The preservation and safety of food products are the two key factors in selecting proper packaging materials. Several environmental factors such as temperature, concentration of O_2/CO_2, humidity, light and external mechanical forces such as vibrations and stresses are the responsible factors for deterioration of food. Microbial attack is another critical reason for the deterioration of packaged food products. Microbial attack not only deteriorates food quality, but also causes food poisoning, leading to food-borne diseases. Nowadays, people are more concerned about smart use of food packaging rather than using preservatives in extending the shelf life of food products. This is because food packaging has a remarkable role in easy transportation and product promotion (description of constituents, uses, way of using the product, marketing and advertisement). However, application of proper packaging material not only plays an outstanding role in extending the shelf life of perishable food products but also helps in easy transportation and exportation of food products. Besides, suitable packaging materials for a specific commodity retain all the food properties and provide wide availability of seasonal food products throughout the year. Depending on the food type and its sensitivity toward different environmental conditions, various packaging materials used are metals, glass, papers or plastics. Metal- and glass-based packaging materials are used for food products that are highly sensitive toward the presence of microbes or external contamination. Specifically the food products that need to be kept in opaque, O_2/CO_2 impermeable conditions such as beer and soft drinks, are packaged in metal can or in glass bottles. Besides, the metal packaging provides longer shelf life for food products and maintains the food quality and nutrition. Furthermore, they have the benefits over their down cycle use with very little changes in their properties [1]. Papers and plastics are the major packaging materials, which contribute toward 70% of packaging material worldwide [2]. Low cost and easy large-scale production are the main reasons for the increased use of paper and plastics in food packaging. Paper-based packages have the advantages of easy recyclability and biodegradability over plastic packaging.

Modification of material properties is necessary to make the material suitable for utilization as perfect food packaging material. As this book is focused on the use of biodegradable plastics as packaging materials, so this branch of knowledge deals with the study of various prospects of packaging systems of food products, which give an indication about the packaging requirements for food products for extending

https://doi.org/10.1515/9783110648034-006

the shelf life of food products. The chapter is also focused on the properties of materials required for food packaging applications. More focused discussion is made on the properties of biodegradable polymers, as research and findings are always trying to replace conventional plastics with biobased plastics due to increasing environmental concerns. However, the selected biobased plastics should have some essential properties for acting as packaging materials. In general, the food packaging materials must possess high barrier properties (oxygen, water vapor, and light), enhanced mechanical properties, thermal properties, rheological properties and oxygen biodegradability. Accordingly, this chapter provides the basic overview of the properties of conventional and biodegradable plastics to be utilized for food packaging applications. Finally, applications of biobased plastics instead of petroleum-based plastics in various categories of foods will be discussed.

6.2 Material properties for food packaging

The conventionally used packaging materials for food products include high-density polyethylene (HDPE), low-density polyethylene (LDPE), medium-density polyethylene (MDPE), polyethylene (PE), polypropylene (PP), polystyrene (PS), polyvinyl chloride (PVC), ethylene vinyl alcohol (EVOH), poly(methyl methacrylate) (PMMA), poly(vinyl alcohol) (PVA), poly(vinyl acetate) (PVAc), polytetrafluoroethylene (PTFE), polyethylene terephthalate (PET) and so on. They have considerable barrier properties against oxygen and water vapor, which provide proper storage for specific food products. However, nondegradable nature of these plastics limits their use as packaging materials. These limitations give rise to the development of photodegradable and biodegradable plastics as packaging materials. However, biodegradable and photodegradable packaging materials should fulfill all the requirements for being a proper packaging material similar to conventional packaging materials. Various biobased-packaging materials for food products include poly(lactic acid) (PLA), poly(3-hydroxybutyrate-*co*-3-hydroxyvalerate) (PHBV), polyhydroxybutyrate (PHB), polycaprolactone (PCL), chitosan, polybutylene succinate (PBS), polybutylene succinate-*co*-adipate (PBSA), cellulose, cellulose acetate, methyl cellulose (MC), cellulose hydrate, cellulose nitrate, thermoplastic starch, nanofibrillated cellulose, nanofirillated chitin and so on. The qualities of an ideal packaging material include lower oxygen and water vapor permeable rate, no light transmission and resistant to mechanical/thermal injuries.

6.2.1 Barrier properties

The barrier properties of a material are essential in food packaging applications. Shelf life of packaged food products mainly depends on O_2/CO_2 permeability and water vapor permeation rate of packaging materials. While most of the food products require impermeable packaging, whereas fruits and vegetables require permeable packaging, which should allow oxygen, carbon dioxide and moisture exchange with external environment to keep them fresh for a longer time. However, the barrier properties of food packaging are determined in terms of permeability, diffusivity and solubility.

6.2.1.1 Water vapor transmission rate (WVTR)

WVTR of packaging material depends on the permeability kinetics of water vapor or moisture through the packaging materials. Depending upon the various requirements for food packaging, sample shape, size and need of accuracy in data, there are several standards such as ASTM E96/E96M-14, ASTM E398-13, ASTM F1249-13, ASTM D3079-94(2009)e1, ISO 15106-1 to 7, ISO 1663:2007(en), ISO 2528:1995(en), ISO 9932:1990(en), DIN 53122-1:2001-08, JIS K 7129:2008, JIS Z 0208:1976, TAPPI T 448 om-09 and TAPPI/ANSI T 464 om-12, which are used to analyze the WVTR or water vapor permeability (WVP). Generally, WVTR is expressed in units g/(m^2·day) or g/(100in.^{2}·day) [3]. Further, the WVP of some of the plastics used in packaging are listed in Table 6.1.

6.2.1.2 Gas transmission rate (GTR)

As stated in the beginning of Section 6.2.1, control on transmission of gases such as oxygen, carbon dioxide and nitrogen through the food packaging is very important in most of the food products. Especially, oxygen transfer rate (OTR) plays a vital role in maintaining the flavor, texture, color, nutrition and shelf life of food products [4]. Similar to WVTR, there are several standards available to evaluate the GTR of polymers; the most common are ASTM F1927-14, ASTM F2622-08(2013), ASTM D3985-05 (2010)e1, ASTM F1307-14, ASTM D1434-82(2009)e1, ASTM F1115-95(2008)e1, ASTM F2476-13, ISO 15105-1 and 2, ISO 2556:1974, ISO 7229:1997, ISO 18369-4:2006, JIS K 7126-1and 2 and DIN 53380-1 to 5. The permeability of oxygen, carbon dioxide and nitrogen of conventional plastics are listed in Table 6.1. GTR primarily depends on the partial pressure of the gas. General methods for GTR calculation are pressure increase method, volume increase method and concentration increase method [5].

6.2.2 Barrier properties of biodegradable polymers

Table 6.2 lists the oxygen permeability and WVP of several nonconventional biodegradable polymers. These biodegradable polymers lack in several properties including

Table 6.1: Permeability of different gases and water in terms of ($cm^{3}\cdot cm/cm^{2}.s.Pa$) through the conventional polymers.

Polymer	H_2O permeability ($\times 10^{13}$)	O_2 permeability ($\times 10^{13}$)	CO_2 permeability ($\times 10^{13}$)	N_2 permeability ($\times 10^{13}$)	References
LDPE	93 (25 °C)	5.18 (25 °C)	21 (25 °C)	1.58 (25 °C)	[4]
HDPE	137 (37.8 °C) ($gm.cm/cm^2.s$)	0.31 (25 °C)	0.28 (25 °C)	0.11 (25 °C)	[6, 7]
PP (isotactic)	38.3 (25 °C)	0.77 (25 °C)	2.37–7.50 (25 °C)	–	[8–10]
PS	0.78–3.9 (WVTR: g. mm/m^2.day)	1.34–1.79	4.49–6.74	0.09–0.11	[11]
PMMA	480 (25 °C)	0.075 (25 °C)	–	–	[4]
PVAc	≈160 (Kgm/Nsec)	0.367 (30 °C)	–	0.066 (below T_g)	[12–14]
PVA	2175–11175 (25 °C)	0.54 (25 °C)	0.90 (25 °C)	–	[15]
PVC	1–2 (23 °C)	0.02–0.09 (23 °C)	–	–	[16]
PTFE	–	3.2 (25 °C)	7.5 (25 °C)	1.0 (25 °C)	[4]
PET	0.5–2 (23 °C)	0.01–0.06 (23 °C)	–	–	[16]

barrier properties, which make them difficult to commercialize. There are several ways to enhance the barrier properties of biodegradable plastics. One way to enhance the barrier properties includes the composite preparation with micro- and nanofillers. Improvement in barrier properties by the addition of inorganic fillers has already been reported in various literatures [17–22]. Biopolymer-based fillers like cellulose, cellulose nanocrystals (CNCs), lignin and chitosan nanoparticles have been also utilized for studying their impact on barrier properties. Sanchez-Garcia and Lagaron [23] have reported the reduction in water permeability of PLA up to 82% and reduction in oxygen permeability up to 90% by using cellulose nanowhiskers. Sanchez-Garcia and coworkers found that low microcellulose fibers composites with PLA, PHBV and PCL thermoplastic biopolymers have improved barrier properties. Miri et al. [25] studied bionanocomposite films of carboxymethyl cellulose (CMC)/starch (ST) polysaccharide matrix reinforced with CNCs and found significant reduction in WVP. Hosseini et al. [26] found 50% reduction in WVP by adding chitosan nanoparticles to fish gelatin-based films.

Table 6.2: Water vapor permeability (WVP) and oxygen permeability (OP) of nonconventional biodegradable polymers.

Polymer	WVP (g·cm/cm^2.s.Pa) at T = 25 °C (P×10^{-13})	OP (cm^3·cm/cm^2.s.Pa) at T = 25 °C (P×10^{-13})
PLA	1.80 [27] 1.42 [28]	0.219 [27] 0.175 [29]
PHBV	1.27 [30]	[a]0.159 [31]
PHB	[b]0.168 [31]	≈0.09 [32]
PCL	3.39 [30]	1.157 [33]
Chitosan	[c]12 [34]	[a]0.0012 [34]
PBS	–	[a]0.302 [35]
PBSA	0.036 [36]	0.359 [37]
Cellulose (cellophane)	142.78 [33]	0.0015 [33]
Cellulose acetate (CA)	[d]124.66 [38]	0.35 [39]
MC	4.75 [40]	5.99 [41]
Cellulose hydrate	18900 [4]	0.0016 [4]
Cellulose nitrate	4720 [4]	1.46 [4]
Starch	[d]9.25 [42]	[c]0.0140 [42]

[a]at 23 °C;
[b]at 38 °Cand 90% RH;
[c]at 23 °Cand 50% RH;
[d]at 20 °C;
[e] at 35 °C.

In situ synthesis of polymer nanocomposites is another way of incorporating fillers in biopolymer matrices to improve the properties. Re et al. [43] prepared PLA nanocomposites by using poly(L-lactide) (PLLA) and poly(D-lactide) (PDLA) grafted on Cloisite 30B nanohybrid fillers, which showed 41% and 46% reduction in oxygen permeability, respectively, when compared to pure PLA. Gain et al. [44] prepared PCL/montmorillonite (MMT) nanocomposites by simple melt blending and in situ polymerization by using MMT clay and observed a reduction in permeability of He, H_2 and CO_2 gases.

Another way to enhance the barrier properties of biodegradable polymers is by using two or more polymers for copolymerization, blending or preparation of multi-layer films. Ligot et al. [45] studied PLA-layered silicate nanocomposites coated with ethyl lactate plasma polymer film and found 53% reduction in oxygen transmission rate. Armentano et al. [46] prepared PLA and PHB blend with lactic acid oligomer plasticizer by extrusion, which showed improvement in both water vapor and oxygen transmission rate.

6.2.3 Mechanical properties

From packaging to delivering food items to the consumer, the food packages have to go through various static and dynamic forces. Mechanical properties of plastics vary in a vast range, but certainly a single polymer does not have all the properties to be able to withstand all such wide range of forces. So the selection of packaging material has to be made depending on the food product condition, the time period, environment conditions, the distance and mode of transport to ship and other marketing requirements. There are various parameters on the basis of which mechanical stability of plastics may be characterized. The most common parameters are (1) tensile strength: the resistance against breaking under tension; (2) impact strength: ability to withstand against sudden applied load; (3) tear strength: resistance against tearing; (4) bursting strength: resistance to failure under pressure; (5) stiffness: resistance to deformation under an applied force and (6) toughness: ability to absorb the energy without fracturing. Coefficient of friction, orientation and shrinkage, crease and flexural resistance are some other parameters that are measured when required for specific applications. Table 6.3 lists the tensile and impact strength of general plastic materials.

6.2.3.1 Mechanical properties of biodegradable polymers

Similar to the fossil-based plastics, biodegradable polymers also display mechanical properties in wide range, which can be observed from Table 6.4. Biopolymers like CNCs are used as fillers to improve the properties of other polymers. On the other hand, biodegradable polymer such as PHB have good tensile strength but is fragile in nature. Similar to PHB most biodegradable polymers have the mechanical stability issues and need further modification in properties to be utilized for packaging applications.

Most of the biodegradable polymers show tensile strength properties in the same range as general plastics, but they are brittle in nature and have less longevity. A number of studies have been carried out to improve the mechanical properties of the biodegradable polymers. The same approaches that are mentioned in the section of the barrier properties, incorporation of micro- and nanofillers and preparing composites, composite formation by in situ polymerization, blending,

Table 6.3: Tensile strength and impact strength of different plastics.

Polymer	Tensile strength (MPa)	Impact strength (J/cm)	References
LDPE	4.1–15.8	No break	[47]
HDPE	10–60	30–200	[48, 49]
PP	29–35	0.3–2	[50]
PS	30–60	19.7	[49, 51]
PMMA	48–76	20 (KJ/m^2)	[52–54]
PVC	56.5	0.91 (ft-lb/in)	[55, 56]
Acrylonitrile butadiene styrene (ABS)	41.8–51.7	5.88–6.95	[47]
PTFE	7–28	1.87 (276 K)	[57]
Nylon 6,6	82.7–90.3 (23 °C)	53–64 (23 °C)	[57]
PC	62.1–65.5	8.50	[49]
Natural rubber	17.24–20.68	–	[47]
Polyurethane	13.79–27.58	–	[47]

copolymerization and multilayer film preparation are used to tailor the mechanical and other properties of the biodegradable polymers. Some techniques significantly improve the properties of biodegradable polymers, which are listed in Table 6.5.

6.2.4 Thermal properties

Mechanical stability of packaging materials in different temperature range is dependent on the thermal properties of the material. It is necessary that from packaging to the delivery of the product, packaging material should not undergo any phase change. Thermal parameters are one of the crucial factors to determine the processing and usable temperature range of a particular polymeric material. The common thermal parameters include glass transition temperature (T_g), crystallization temperature (T_c), melting temperature (T_m)/decomposition temperature (T_d) and so on. Heat deflection temperature (HDT), heat capacity (C) and heat of fusion (ΔH_f) are some other parameters, which also have to be taken into account during processing and application. Table 6.6 lists these thermal parameters of general plastics.

Table 6.4: Tensile strength and impact strength of different biodegradable plastics.

Polymer	Tensile strength (MPa)	Impact strength (J/cm)
PLA	35.8 [58]	0.20 [59]
PHBV	37.8 [60]	0.20 [60]
PHB	25.4 [61]	0.22 [62]
PCL	15.7 [63]	1.20 [64]
Chitosan	8.25 [65]	20 (MPa) [66]
PBS	30 [67]	0.85 (J/cm^2) [67]
PBSA	10 – 15 [68]	0.003 (J/cm^2) [68]
PBAT	20.2 [69]	No Break [69]
CA	26 [49]	1.73 [70]
Nanofibrillated cellulose	118 [71], 242 [72]	–
Nanofibrillated chitin	132 [71]	118 [71], 242 [72]
Starch	1.4–14.2 [73, 74] (varies with source of extraction)	–

Table 6.5: Improvement in the different mechanical properties of biodegradable polymers by using specific techniques.

Base polymer	Technique to control the mechanical properties	Materials used for improvement	Mechanical property	% Increment	References
PBS	Composite formation	Graphene oxide	Tensile strength	53	[75]
			Young's modulus	70	
			% Elongation	≈2	
PCL	Blending	PLA	Yield stress	Improved with increase in PLA fraction	[76, 77]
			Young's modulus		

Table 6.5 (continued)

Base polymer	Technique to control the mechanical properties	Materials used for improvement	Mechanical property	% Increment	References
PHB	Composite formation	Rice Hull ash (decreased the brittle nature of PHB)	Impact resistance	≈20	[78]
			Hardness	≈13	
			Tensile strength	−21	
			Modulus of elasticity	−5	
Cellulose cardboard	Bilayer formation	PHB	Young's modulus (solvent casting)	40	[79]
			Young's modulus (compression molding)	65	
PCL	Nanocomposite formation	Modified Bagasse whisker	Young's modulus	77	[80]
Guar gum	Nanocomposite formation	Nanofil 116	Tensile strength	101	[81]
			Young's modulus	5,400	
			% Elongation	1	
		Cloisite 20A	Tensile strength	41	
			Young's modulus	800	
			% Elongation	4.5	
PBS	In situ polymerization	Silica	Tensile strength	46	[82]
			% Elongation	418	

Table 6.5 (continued)

Base polymer	Technique to control the mechanical properties	Materials used for improvement	Mechanical property	% Increment	References
Epoxidized soybean oil	In situ synthesis	Multiwalled carbon nanotubes	Scratch hardness	130	[83]
			Tensile strength	390	
PLA	Nanocomposite formation	TiO_2 nanoparticles	Tensile strength	80	[84]
			Young's modulus	129	
			% Elongation	16.5	
CMC	Composite formation	Chitosan-PMMA nanoparticles	Tensile strength	540	[85]
			% Elongation	2.1	
CA	Nanocomposite formation	Acetylated cellulose nanowhiskers	Tensile strength	9	[86]
			Young's modulus	39	
			Strain at break	44	
PHB	Blending	PLA	Tensile strength	PLA-PHB 75–25 wt% had better values compared to both neat polymers	[87]
			% Elongation		
PHB	Bilayer formation	Zein	Elastic modulus	18	[88]
			Tensile strength	8	
PBS	Photografting polymerization	Acrylic acid	Young's modulus	4	[89]

Table 6.6: Thermal parameters of general plastics.

Polymer	Glass transition temperature T_g (°C)	Melting temperature T_m (°C)	Decomposition temperature T_d (°C)	References
LDPE	−133 to −103	105–115	–	[57]
HDPE	−95	130–146	360–390	[90]
PP	−18	160–208	336–366	[90]
PS	97	–	318–348	[90]
PMMA	104	–	226–256	[90]
PVC	80	–	≈250	[90]
PTFE	−20	327	424–513	[90]
Nylon 6,6	47–57	264	–	[57]
PC	145	–	≈327	[90]
Polyurethane	22–29	203–225	–	[57]
PEEK	120	340	–	[90]
PAN	97	320	–	[57]
PVA	85	265	–	[57]
PVAc	28–31	175	–	[57]
PET	69–115	245–265	–	[57]

6.2.4.1 Thermal properties of biodegradable polymers

Table 6.7 lists the thermal properties of few biodegradable polymers. Most of the biodegradable polymers have low T_g, which makes their use difficult for normal packaging applications. In addition, most of the biodegradable polymers have very small difference between T_m and T_d, which makes their processing very difficult.

The usability and processing challenges can be overcome by tuning the thermal properties of biodegradable polymers. Composite formation by adding fillers, blending, copolymerization and multilayer film forming is the major technique that has been utilized for tuning the thermal properties. Chemical modifications such as crosslinking and use of additives like plasticizers are also common in practice to control the thermal parameters for better processability. Table 6.8 reviews several work available in literature where the thermal properties of biodegradable polymers have been optimized for better processability and different applications.

Table 6.7: Thermal properties of biodegradable polymers.

Polymer	Glass transition temperature T_g (°C)	Melting temperature T_m (°C)	Decomposition temperature T_d (°C)	References
PLA	55–56	120–170	above 200 (through intra- and intermolecularester exchange, cis-elimination, radical and concerted non-radical reactions)	[91]
PHBV	-5 to 20	108	–	[92]
PHB	55	180	–	[92]
PCL	-60	60–65	–	[92]
Chitosan	130	–	285–305	[93]
PBS	-45 to -45	90–120	–	[92]
PBSA	-41.2	93.3	–	[94]
PBAT	-33.2 to -29	124–126	–	[95–97]
CA	189.1	221.8	–	[98]
CMC	25–65	–	278	[99]
Starch	62.8	170–190	–	[100, 101]

6.2.5 Morphological properties

Morphology defines the behavior of polymeric materials and it influences on the thermal, mechanical, barrier and optical properties. Furthermore, several parameters such as percentage crystallinity, crystal size and shape, crystal arrangements and distribution can affect the properties of polymers. In case of biodegradable polymers, morphological parameters become more crucial as they can significantly vary the properties of these polymers, which may help to tailor their properties for different applications. For example, if the crystalline part is selectively taken out from cellulosic biomass, the crystalline cellulose shows very good mechanical properties. Similar is the case of PLA; higher crystalline PLA shows improved thermal and mechanical properties. Table 6.9 summarizes few examples of such work, where properties of biodegradable polymers have been improved by tailoring the morphology of the polymer.

Table 6.8: Thermal properties modification of biodegradable polymers.

Base polymer	Modification technique	Modifier	Thermal property	Increment in °C	References
PLA	In situ polymerization	Surface-hydroxylated MgO	T_d (onset)	27.2	[102]
PBSA	Nanocomposite formation	Cloisite-30B Clay	T_d (onset)	39.6	[103]
Chitosan	Crosslinking	4,4′(Hexafluoroisopropylidene)diphthalic anhydride	T_d (max)	60	[104]
		4,4′-Oxydiphthalic dianhydride		43	
		Benzophenone-3,3′,4,4′-tetracarboxylic dianhydride		47	
		Pyromellitic dianhydride		43	
Cellulose	Blending	PCL	T_d (initiation)	35	[105]
PBS	Nanocomposite formation	Carbon black	HDT	7	[106]
PHBV	Nanocomposite formation	Functionalized graphene sheets	T_g	7	[107]
Gallic acid epoxy	Cationic ring opening reaction	Multiwalled carbon nanotubes	T_d (initiation)	10.9	[108]
			T_d (half)	86.6	
Hydroxypropylmethylcellulose	Plasticizing	Suberin fatty acids	T_g	−8.6	[109]
		PEG		−30.1	
		Suberin fatty acids + PEG		−25.5	

(continued)

Table 6.8 (continued)

Base polymer	**Modification technique**	**Modifier**	**Thermal property**	**Increment in °C**	**References**
PCL–PGA blend	Nanocomposite formation by compression molding	dsDNA wrapped single-walled carbon nanotubes	T_g	8.8	[110]
Poly(lactide-*co*-glycolide) polymerized by hydroxyapatite	Copolymerization and crosslinking	Succinic anhydride, *N*-hydroxysuccinimide, dicyclohexylcarbodiimide, calf skin collagen	T_d	~40	[111]
PLLA	Composite formation by melt compounding	Cork	T_d (max)	−16.4	[112]
			T_g	1.1	
PHBV			T_d (max)	13	
			T_g	5.2	
PCL			T_d (max)	−127.2	
Starch-PCL			T_d (max)	5.2	

Table 6.9: Effect of change in morphology on different properties of biodegradable polymers.

Base polymer	Material or method for changing the morphology	Property	Parameters	% Improvement	References
PLA	Annealing	Barrier properties	OP	≈80	[113]
			WVP	≈70	
PLA–PBS blend	Intercalation and exfoliation of organoclay	Mechanical properties	Young's modulus	41	[114]
		Barrier properties	OP	40	
			WVP	50	
PDLLA	CNC-driven crystallization	Mechanical properties	Young's modulus	40	[115]
		Thermal properties	T_m	46	
			T_g	21	
		Thermomechanical	Storage modulus	≈2,820	
PBS	Nucleation by halloysite nanotube	Mechanical properties	Flexure modulus	13.4	[116]
			Tensile modulus	16.2	
			Impact strength	26	
CA	Exfoliation and dispersion of Closite30B used with other plasticizer and compatibilizer	Mechanical properties	Tensile strength	20	[117]
			Tensile modulus	68	
			Flexural strength	20	
			Flexural modulus	25	
PLLA–PCL blend	Nucleation	Mechanical properties	Impact strength	≈850	[118]

Table 6.9 (continued)

Base polymer	Material or method for changing the morphology	Property	Parameters	% Improvement	References
PLA	Formation of blend by converting nanofibrillar PBS to nanolaminar	Barrier properties	OP	86	[119]
		Mechanical properties	Yield strength	28	
			Young's modulus	≈25	
PBS		Barrier properties	OP	67	
Amorphous cellulose film (ACF)	ACF formed form cellulose solution in lithium chloride (8wt%)/*N*, *N*dimethyl-acetamide by regeneration with acetone	Mechanical properties (compared to cellophane)	% Elongation	4	[120]
			Maximum tensile stress	16	
		Optical properties	Transparency	Better than cellophane	
PHB	Alkyl functionalized and acid-treated MWCNT as nucleating agents for crystallization	Thermomechanical properties	Storage modulus (at −20 °C)	75–85	[121]
PLA	Ethylene bis-stearamide-treated organo-modified clay, halloysite nanotubes and silica as nucleating agents	Mechanical properties	Young's modulus (max)	47	[122]
			Impact resistance (max)	110	
CA	Multilayer film forming with LDH and thermal annealing	Barrier properties	OP	Below the detection level	[123]

6.2.6 Optical properties

Transparency, haziness, gloss, opacity and refractive index are the parameters that define the optical properties of polymeric materials. Transparent and glossy packaging is preferred over hazy and opaque packaging. In the case of food, transparent packaging is even more preferable, specifically for the food that has shorter life time as the customers can visually identify the quality of food packaged. Transparency is mainly dependent on the morphology of polymers and can be tuned by controlling the morphology. As the crystallinity of polymers increases, their transparency decreases and the transparent polymers do not absorb all wavelength of visible light. As the crystallinity increases, the scattering of visible light increases, because the crystal structure has the size in the range of wavelength of visible light [124]. An increase in scattering results in the reduction of transparency. Various reports are available where optical properties of biodegradable polymers have been controlled by varying the morphology, by changing the crystallinity, processing conditions or by addition of fillers. Herrera et al. [125] studied the effect of adding cellulose and chitin nanocrystals to the plasticized PLA by extrusion and compression molding with two cooling rates. Faster cooling rate produces material with more amorphous nature and improved transparency relative to the slowly cooled one. Oliveira et al. [126] prepared CMC–laponite clay nanocomposite membrane. The uniform dispersion of clay resulted in the improved transparency of nanocomposite film of CMC–laponite compared to the CMC film. Ahmad et al. [127] studied glycerol-plasticized fish gelatin-rice flour composite membranes, which showed improved transparency and excellent light barrier characteristics compared to the fish gelatin film.

6.2.7 Rheological properties

The response of polymers in terms of flow properties on application of external shear force defines the behavior of the polymer fluid. Almost all polymers are non-Newtonian in nature and have viscoelastic behavior. The viscoelastic nature of polymers is determined by the rheological parameters. Rheological parameters are the guiding factors in determining the working force, time and temperature ranges of polymeric materials undergoing continuous static or dynamic force loading and/or temperature variable conditions. Different chemical treatments, fillers, additives, blends and several other modifications may influence the rheological properties of polymers on varying the processing conditions. Few of such techniques used for biopolymers and their impact on the rheological properties are listed in Table 6.10.

Table 6.10: Different techniques to modulate the rheological properties of biopolymers and their impact.

Base polymer	Technique	Material used to vary the rheology	Changes in rheological parameter	References
PLA–PBSA blend	Use of chain extender and clay composite formation	Epoxy chain extender and Cloisite-30B clay	Increase in η and higher G' compared to the blend	[128]
PBS	Composite formation	Nano-$CaCO_3$	Improvement in E' and E'', viscous and shear thinning effect was observed for higher frequencies	[129]
CMC-starch composite	Nanocomposite formation	CNC	Shear thinning effect and film-forming properties at room temperature are observed	[25]
Whey protein isolate	Blending	Chicken skin gelatin	High G', stable and frequency-independent gel formation	[130]
Maize flour	Blending	Starch and zein	Increment in elongational viscosity	[131]

6.2.8 Biodegradability

The breakdown of a material in vitro or in vivo by catalytic activity of enzymes is defined as biodegradation [132]. Biodegradable polymers mainly contain hydrolysable and/or oxidizable groups in main chains and biodegradation takes place by hydrolysis and oxidation reactions [133]. Polyesters derived from microorganisms, natural polysaccharides and synthetic polymers (aliphatic polymers) are the three major categories of biodegradable polymers [134]. The degradability of plastics has become a major concern with increasing plastic waste due to nonbiodegradability of conventional plastics, which drives toward the requirement of substitutes that are biodegradable in nature. The parameters affecting the biodegradation mainly include temperature, light, humidity, nutrients, pH, oxygen, molecular form, morphology and number and type of microbes. Large-scale composting of these polymers takes about 6–12 weeks for complete degradation of bioplastics to water and carbon dioxide [135, 136]. Biodegradation can occur in both aerobic and anaerobic environment. Generally, biodegradation is a two-step process, the first step being depolymerization in which longer polymer chains decompose into shorter chains and the second step is mineralization in which small oligomer chains transport into cells of microorganisms, where they get mineralized and produce CO_2, CH_4, N_2, H_2, water, salts, minerals and biomass [137].

6.3 Petroleum and biobased food packaging materials

From commodity to commodity, food products have a wide range of shelf life at various storage conditions (temperature, pressure, relative humidity, O_2/CO_2 concentration, etc.). Some of the food products such as meat, milk, climacteric fruits and vegetables have a very short shelf life of around 2–3 days at refrigeration temperature due to their poor barrier against environment. On the other hand, careful handling of some products such as onion and potato can lead to enhanced shelf life and keeping them in shelves at room temperature can retain all the food properties, which prevent them from deterioration. In contrast, some food do not need any treatment or packaging system for maintaining their quality as their own outer cover acts as a barrier against surroundings. Such food products degrade only due to careless handling. This discussion classifies the food products into three groups as given below:
(1) Perishable food
(2) Semiperishable food
(3) Nonperishable food

Perishable food: Perishable food are those food that deteriorate very fast if not stored properly under refrigeration conditions. Although proper storage extends their shelf life, a proper packaging to protect them from microorganisms, oxygen and water vapor is a must. This category of food mainly includes dairy products, meat and meat products, poultry, sea foods, fruits and vegetables and cooked foods.

Semiperishable food: Semiperishable food are those that have better shelf life than perishable food at room temperature but require proper storage and careful handling. Examples includes potatoes, onion and so on.

Nonperishable food: Nonperishable food are those that do not get deteriorated easily unless handled carelessly and include mainly grains, flour, dry products and so on.

6.3.1 Packaging materials for milk and milk products

6.3.1.1 Importance of milk and milk products

Milk provides all health beneficial nutrients required for growth of human body. Indeed, it is known as a type of complete food because it contains all beneficial nutrients [138]. Milk varies in its constituents according to its source and place such as type of animal (buffalo milk, cattle milk and goat milk), type of fodder taken, type of milking (hand milking or machine milking), type of place, environmental conditions and so on. Among all milk types, goat milk is known as the naturally homogenized milk, which is a great replacement for mother's milk for the infants. However, it is unacceptable many a times due to its bad odor. The quality rating of milk is done based on its fat and SNF (solid not fat) content. Buffalo milk has higher fat content than cow milk. Table 6.11 describes a range of various constituents of milk. In addition, the by-products of milk such as *ghee*, yogurt, cream, ice-cream, *khoa, rabri, kheer* and so on are obtained by varying the composition of milk [138]. Milk is one of the primary food products of India and the country ranks high in milk production due to the widespread industrial production of milk and milk products. One of the most renowned industries that brought the white revolution in India is Amul, located in the state of Gujarat. But the main drawback of milk products is their shelf life, which is directly related to the processing and packaging systems. Milk, having very small shelf life, requires proper processing/packaging to protect it from environmental impacts. Milk and milk products are very light sensitive and oxygen sensitive and are also very prone to microbial attack. The deterioration of milk can be identified by its off-flavor, off-taste, rancid, oily, fishy flavors and so on. The packaging of milk and milk products includes various petroleum-based, biobased packaging, glass bottles, tin cans and others, which have many advantages over each other. Biobased packaging is the demand of the present age for milk packaging due to its biodegradable nature.

Table 6.11: Composition of milk.

Milk constituents	Quantity present (%)
Water	85–88.7
Protein	3–4
Fat	3–6
Total solid	11.4–14.5
Sugar	5

6.3.1.2 Present and future prospects of milk packaging

The long-established packaging materials for milk are glass bottles, coated paper board cartons and so on. Glass bottles are largely preferred for milk packaging as they have all the essential qualities for retaining the properties of milk, but the brittle nature of glass materials restricts their use in safe storage and transportation. The use of plastic packaging materials overcomes the shortcomings of glass bottles and the widely used plastic packaging materials include PET and coextruded HDPE, which are nondegradable and impose a serious threat to the environment. The use of packaging systems for any product depends upon the oxygen and water vapor transmission rate of the packaging systems. But resource limitations and increasing environmental concerns have led to the development of biobased plastics (bioplastics) as packaging materials having competent oxygen and water vapor transmission rate. Besides, PLA can replace conventional petroleum-based plastics such as LDPE, linear low-density polyethylene (LLDPE), PP and so on. Along with packaging applications, biobased materials can be used as filler materials. The used biobased packaging for milk and milk products are given in Table 6.12. Nowadays, packaging materials are also made from whey protein of milk.

Table 6.12: Present and future prospects of milk packaging systems.

Milk and milk product	Applied packaging systems
Milk	PET bottles
Milk	Coextruded HDPE bottles
Dairy drink	PLA bottles
Milk chocolates	Corn starch trays
Cheese	PHA, PLA
Yoghurt	PLA
Milk	Tetra pack
Cream	Tetra pack

Tetra packs are a kind of packaging systems used for milk or fruit juice. Tetra packs have six layers, each layer having its own significance. Figure 6.1 shows various layers of tetra packs. PE layer provides moisture protection and enables paper board to stick to the aluminum foil. Aluminum foil protects milk and milk products from oxygen and light along with retaining the flavor of food. It is difficult to separate aluminum foil and plastic material. Only 5% of the aluminum foil can be separated from tetra packs and recycled. This leads to significant amount of material wastage. But now, the researchers have come up with the idea of using EVOH, which has excellent gas barrier properties to prevent deterioration of food by oxygen. Using EVOH in place of aluminum will reduce the need for separation of aluminum from plastic and will ultimately reduce the

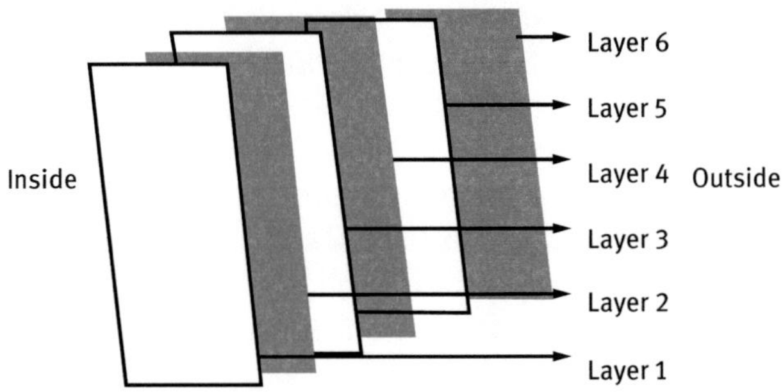

Figure 6.1: Layers of tetra pack: Layer 1: PE, layer 2: PE, layer 3: aluminum foil, layer 4: PE, layer 5: paper board and layer 6: PE.

amount of waste going to landfill. Paper board in tetra pack gives strength and smoothness to the printing surface. The essential properties of material used for milk packaging are oxygen barrier properties, light barrier properties and resistance to microbial attack. Nowadays, caps of tetra packs are replaced with biobased plastics. The layers of tetra packs can be replaced with biobased plastics. Instead of PP/PS, starch as filler materials in PP/PS can be used, which have better degradable nature.

6.3.2 Packaging materials for meat and poultry

6.3.2.1 Importance of meat and poultry

Meat is the flesh of domesticated animals such as cattle, sheep, pigs and so on. On the other hand, poultry are the domesticated birds that include chickens, duck, geese, quails and turkeys. The consumable parts include flesh, organs and eggs. Both meat and poultry are the outstanding sources of protein and contain about 15–20% and 18–25% protein, respectively [138]. Other constituents present are essential amino acids, little fat, minerals, vitamins and so on. Poultry is a good source of vitamin B. However, these foods are perishable and so it is essential to store them properly. The red color of meat is due to the presence of myoglobin pigment, which gets oxidized and forms metmyoglobin when it comes in contact with air, which the consumers are unwilling to accept. This leads to the need of proper packaging materials for meat and poultry so as to protect them from environmental effects.

6.3.2.2 Packaging systems for meat and poultry

Meat and meat products are prone to oxidation and microbial attack and water-absorbing nature of meat makes the product juicy and also changes the taste of the product, which is not acceptable. Since decades people have been using PE bags, trays of LDPE, HDPE for packing of meat. However replacing petroleum-based packaging with the biobased packaging is the need of the hour. Table 6.13 describes about the petroleum-based and biobased packaging systems for meat and poultry products.

6.3.3 Packaging materials for fruits and vegetables

6.3.3.1 Importance of fruits and vegetables

Fruits are formed naturally from the ovary of flowering plants and have seeds within it. Most widely available fruits include apple, mango, banana, strawberry, blueberry and so on. On the other hand, vegetables are considered as plants and eatable part of plants that serve as the main course of meal. Vegetables include onions, potatoes, cucurbit, cabbage, cauliflower and so on. However, tomatoes fall under the categories of both fruits and vegetables. Most of the fruits and vegetables have a hard covering around, which protects them from the surrounding.

Table 6.13: Packaging systems for meat and poultry products.

Meat products	Packaging system	Discussion	References
Fresh chicken breast meat	LDPE/PA/LDPE	Shelf life extension	[139]
Meat	PS	High speed production of PS foam meat packaging trays	[140]
Poultry	PS overlapped with PVC or polyoleifin film over trays	Shelf life extension	[141]
Raw fresh meat	PVC film	Moisture retention	[142]
Beef	PP film	Stability of myoglobin	[143]
Sliced fresh red meat	Chitosan incorporated LDPE film	Shelf life extension	[144]
Sliced Ham	Cellulose film/PE + nisin	Decrease in lactic acid bacteria	[145]

Fruits contain higher amounts of water (80–90%) and other constituents such as proteins, fats, lipids, vitamins, minerals and so on. Several fruits also contain bioactive agents, which have the essential properties such as antioxidant, antimicrobial, anticancer, antibacterial, antidiabetic and so on. The fruits such as pomegranate, apple, strawberries, fig and so on are rich sources of bioactive compounds and are stored in an environment so as to protect them from insects, microorganism and other factors.

6.3.3.2 Packaging systems for fruits and vegetables

Traditional packaging systems used for fruits and vegetables include LDPE, HDPE, PS, PVC, PP, MDPE, tin can and so on. Table 6.14 describes the impact of various polymers on different fruits and vegetables. The packaging materials vary in their oxygen and water vapor permeability rate according to the storage conditions (temperature, storage period and relative humidity). The gas concentrations inside the packaged foods can be determined by using PBI dansensor that gives the information about OTR/WVTR properties of packaging materials. Nowadays, modified atmospheric packaging (MAP), controlled atmospheric packaging (CAP), vacuum packaging, active packaging and smart packaging are gaining increasing attention as they provide environmental conditions that increase the shelf life of commodities. However, specific packaging materials also play an important role in extending the shelf life of specific fruits and vegetables according to their respiration rates and permeability of their packaging materials.

Knowledge of various properties of plastics (petroleum-based and biobased plastics) materials is essential to identify the most suitable material for food-packaging application. It is very important to understand the dynamic behavior of polymers and their stability in different environmental conditions. In this part of the chapter, the required properties of polymers for food-packaging applications have been summarized, with significant focus on packaging systems of some perishable food products. It is important for packaging materials to retain the properties of perishable food materials so as to extend their shelf life for wider acceptance. The polymers for packaging applications have been derived from various renewable sources such as milk protein, other protein sources, cellulose sources and so on. However, significant developments are required for developing biobased polymeric materials for packaging applications that can gain commercial and environmental acceptance so as to completely replace conventional petroleum-based plastics.

Table 6.14: Conventional packaging materials for fruits and vegetables.

Commodity	Polymeric materials	Discussion	References
Chiku (*Achras Sapota L*)	LDPE at 5, 10, 15 °C	Maintain texture and quality	[146]
Apple	HDPE and 1 MCP sorption capacity	No weight gain in the packaging	[147]
Mango milk powder, mango milk drink	HDPE, tin can, metalized polyester	Quality retained	[148]
Blanched mango pulp	Polyetlrylene, polypropylene, paper-Al foil-polyethylene laminate at 12 °C	Shelf life extension and quality retained	[149]
Banana cultivar Sucrier	PVC stored for 6–7 days	Decrease dopamine content	[150]
Strawberry syrup	Glass bottles, PVC, PET	Aroma retention	[151]
Guava cv Kumagai	Multilayer coextruded polyolephine film with selective permeability	Shelf life extension	[152]
Guava cv Kumagai	LDPE film + minerals	Shelf life extension	[152]
Guava cv Kumagai	Polyolephine film	Shelf life extension	[152]
Sapota	MAP (LDPE, HDPE, PVC, PP, PS)	Decrease respiration rate	[153]
Blueberries	PLA at 10 °C	Extends shelf life	[154]
Fresh cut green salad	Thermoformed oriented polylactic acid	Retain quality	[155]
Cucumber and bell pepper	Chitosan	Reduction in water loss and maintain quality	[156]
Minimally processed strawberry	Chitosan	Preservation, increase antimicrobial activity	[157]
Litchi	Chitosan	Lowers enzymatic browning and decay during storage	[158]

References

[1] Benefits of Metal Packaging, Washington DC, North America Packaging Alliance Inc, 2015. Accessed June 6, 2015, at. http://metal-pack.org/docs/pdf/00043919.PDF.
[2] Kim, YT., Min, B., & Kim, KW. General characteristics of packaging materials for food system, Han JH., Innovations in Food Packaging, 2nd ed, Texas, USA, Academic Press, 2014, 13–36.
[3] Yam, KL. The Wiley Encyclopedia of Packaging Technology, New Jersey, John Wiley & Sons, 2009.
[4] Piringer, OG., & Baner, AL. Plastic packaging materials for food: barrier function, mass transport, quality assurance, and legislation, Weinheim, WILEY-VCH Verlag GmbH, 2000.
[5] Paine, FA., & Paine, HY. A Handbook of Food Packaging, Dordrecht, Springer-Science +Business Media, 1992.
[6] Michaels, AS., & Bixler, HJ. Solubility of gases in polyethylene. J Polym Sci 1961, 50, 393–412.
[7] Kamal, MR., Jinnah, IA., & Utracki, LA. Permeability of oxygen and water vapor through polyethylene/polyamide films. Polym Eng Sci 1984, 24, 1337–1347.
[8] Myers, AW., Stannett, V., & Szwarc, M. The permeability of polypropylene to gases and vapors. J Polym Sci 1959, 35, 285–288.
[9] Taraiya, AK., Orchard, GA., & Ward, IM. The diffusion of oxygen through uniaxially and biaxially drawn polypropylene. J Appl Polym Sci 1990, 41, 1659–1671.
[10] Naito, Y., Mizoguchi, K., Terada, K., & Kamiya, Y. The effect of pressure on gas permeation through semicrystalline polymers above the glass transition temperature. J Polym Sci B 1991, 29, 457–462.
[11] McKeen, LW. Permeability properties of plastics and elastomers, Oxford, William Andrew, 2016.
[12] Mark, HF. Encyclopedia of Polymer Science and Technology, New York, Wiley-Interscience, 1987.
[13] Meares, P. The diffusion of gases through polyvinyl acetate1. J Am Chem Soc 1954, 76, 3415–3422.
[14] Chien, AT., & Lin, KF. Morphology and permeability of exfoliated PVAc-MMT nanocomposite films cast from soap-free emulsion-polymerized latices. J Polym Sci A 2007, 45, 5583–5589.
[15] Mark, HF., McKetta, JJ., & Othmer, DF. Kirk-Othmer encyclopedia of chemical technology, New York, Wiley-Interscience, 1963.
[16] Lange, J., & Wyser, Y. Recent innovations in barrier technologies for plastic packaging—A review. Packag Technol Sci 2003, 16, 149–158.
[17] Rhim, JW. Effect of clay contents on mechanical and water vapor barrier properties of agar-based nanocomposite films. Carbohydr Polym 2011, 86, 691–699.
[18] Müller, CM., Laurindo, JB., & Yamashita, F. Effect of nanoclay incorporation method on mechanical and water vapor barrier properties of starch-based films. Ind Crops Prod 2011, 33, 605–610.
[19] Maksimov, RD., Lagzdins, A., Lilichenko, N., & Plume, E. Mechanical properties and water vapor permeability of starch/montmorillonite nanocomposites. Polym Eng Sci 2009, 49, 2421–2429.
[20] Famá, L., Rojo, PG., Bernal, C., & Goyanes, S. Biodegradable starch based nanocomposites with low water vapor permeability and high storage modulus. Carbohydr Polym 2012, 87, 1989–1993.
[21] Alves, VD., Castelló, R., Ferreira, AR., Costa, N., Fonseca, IM., & Coelhoso, IM. Barrier properties of carrageenan/pectin biodegradable composite films. Procedia Food Sci 2011, 1, 240–245.

[22] Kumar, P., Sandeep, KP., Alavi, S., Truong, VD., & Gorga, RE. Preparation and characterization of bio-nanocomposite films based on soy protein isolate and montmorillonite using melt extrusion. J Food Eng 2010, 100, 480–489.
[23] Sanchez-Garcia, MD., & Lagaron, JM. On the use of plant cellulose nanowhiskers to enhance the barrier properties of polylactic acid. Cellulose 2010, 17, 987–1004.
[24] Sanchez-Garcia MD., Lopez-Rubio A., Lagaron JM. Natural micro and nanobiocomposites with enhanced barrier properties and novel functionalities for food biopackaging applications. Trends in Food Science & Technology 2010, 21, 528-536.
[25] El Miri, N., Abdelouahdi, K., Barakat, A., Zahouily, M., Fihri, A., Solhy, A., & El Achaby, M. Bio-nanocomposite films reinforced with cellulose nanocrystals: rheology of film-forming solutions, transparency, water vapor barrier and tensile properties of films. Carbohydr Polym 2015, 129, 156–167.
[26] Hosseini, SF., Rezaei, M., Zandi, M., & Farahmandghavi, F. Fabrication of bio-nanocomposite films based on fish gelatin reinforced with chitosan nanoparticles. Food Hydrocoll 2015, 44, 172–182.
[27] Rhim, JW., Hong, SI., & Ha, CS. Tensile, water vapor barrier and antimicrobial properties of PLA/nanoclay composite films. LWT Food Sci Technol 2009, 42, 612–617.
[28] Dhar, P., Tarafder, D., Kumar, A., & Katiyar, V. Effect of cellulose nanocrystal polymorphs on mechanical, barrier and thermal properties of poly (lactic acid) based bionanocomposites. RSC Adv 2015, 5, 60426–60440.
[29] Petersen, K., Nielsen, PV., & Olsen, MB. Physical and mechanical properties of biobased materials starch, polylactate and polyhydroxybutyrate. Starch-Stärke 2001, 53, 356–361.
[30] Sanchez-Garcia, MD., Gimenez, E., & Lagarón, JM. Morphology and barrier properties of solvent cast composites of thermoplastic biopolymers and purified cellulose fibers. Carbohydr Polym 2008, 71, 235–244.
[31] Lagarón, JM. Multifunctional and nanoreinforced polymers for food packaging, Lagarón JM., ed, Multifunctional and Nanoreinforced Polymers for Food Packaging, Cambridge, Woodhead Publishing, 2011, 1–28.
[32] Díez-Pascual, A., & Díez-Vicente, A. Poly (3-hydroxybutyrate)/ZnO bionanocomposites with improved mechanical, barrier and antibacterial properties. Int J Mol Sci 2014, 15, 10950–10973.
[33] McKeen, LW. Permeability properties of plastics and elastomers, Cambridge, Elsevier Inc., 2012.
[34] Boesel, LF. Effect of plasticizers on the barrier and mechanical properties of biomimetic composites of chitosan and clay. Carbohydr Polym 2015, 115, 356–363.
[35] Siracusa, V., Lotti, N., Munari, A., & Dalla Rosa, M. Poly (butylene succinate) and poly (butylene succinate-co-adipate) for food packaging applications: gas barrier properties after stressed treatments. Polym Degrad Stab 2015, 119, 35–45.
[36] Berthet, MA., Angellier-Coussy, H., Guillard, V., & Gontard, N. Vegetal fiber-based biocomposites: which stakes for food packaging applications?. J Appl Polym Sci 2016, 133, 42528–42546.
[37] Kathuria, A. Biodegradable Polymer Nanocomposites from Poly (alkylene Decarboxylate) and Organically Modified Montmorillonite Clay, Michigan State University, School of Packaging, 2007.
[38] Kabiri, R., & Namazi, H. Nanocrystalline cellulose acetate (NCCA)/graphene oxide (GO) nanocomposites with enhanced mechanical properties and barrier against water vapor. Cellulose 2014, 21, 3527–3539.
[39] Nakai, Y., Yoshimizu, H., & Tsujita, Y. Enhanced gas permeability of cellulose acetate membranes under microwave irradiation. J Membrane Sci 2005, 256, 72–77.

[40] Ahmadi, E., Sareminezhad, S., & Azizi, MH. The effect of ultrasound treatment on some properties of methylcellulose films. Food Hydrocoll 2011, 25, 1399–1401.
[41] Ayranci, E., & Tunc, S. A method for the measurement of the oxygen permeability and the development of edible films to reduce the rate of oxidative reactions in fresh foods. Food Chem 2003, 80, 423–431.
[42] Ghasemlou, M., Aliheidari, N., Fahmi, R., Shojaee-Aliabadi, S., Keshavarz, B., Cran, MJ., & Khaksar, R. Physical, mechanical and barrier properties of corn starch films incorporated with plant essential oils. Carbohydr Polym 2013, 98, 1117–1126.
[43] Re, GL., Benali, S., Habibi, Y., Raquez, JM., & Dubois, P. Stereocomplexed PLA nanocomposites: from in situ polymerization to materials properties. Eur Polym J 2014, 54, 138–150.
[44] Gain, O., Espuche, E., Pollet, E., Alexandre, M., & Dubois, P. Gas barrier properties of poly (ε-caprolactone)/clay nanocomposites: influence of the morphology and polymer/clay interactions. J Polym Sci B 2005, 43, 205–214.
[45] Ligot, S., Benali, S., Ramy-Ratiarison, R., Murariu, M., Snyders, R., & Dubois, P. Mechanical, optical and barrier properties of PLA-layered silicate nanocomposites coated with organic plasma polymer thin films. Mater Sci Eng Adv Res 2015, 1, 20–30.
[46] Armentano, I., Fortunati, E., Burgos, N., Dominici, F., Luzi, F., Fiori, S., Jiménez, A., Yoon, K., Ahn, J., Kang, S., & Kenny, JM. Processing and characterization of plasticized PLA/PHB blends for biodegradable multiphase systems. Express Polym Lett 2015, 9, 583–597.
[47] Chanda, M., & Roy, SK. Industrial polymers, specialty polymers, and their applications, Boca Raton, CRC press, 2009.
[48] Kennedy, MA., Peacock, AJ., & Mandelkern, L. Tensile properties of crystalline polymers: linear polyethylene. Macromolecules 1994, 27, 5297–5310.
[49] Mark, JE. Physical Properties of Polymers Handbook, Woodbury, New York, AIP Press, 1996.
[50] Brandrup, J., Immergut, EH., & Grulke, EA. Polymer Handbook, New Jersey, John Wiley & Sons, Inc, 1999.
[51] Van Krevelen, DW. Properties of polymers: their correlation with chemical structure; their numerical estimation and prediction from additive group contributions, Amsterdam, Elsevier, 2009.
[52] Thompson, LF., Willson, CG., & Fréchet, JM. Materials for microlithography: radiation-sensitive polymers, Washington, D.C., American Chemical Society,, 1984.
[53] Billmeyer, FW. Textbook of Polymer Science, New York, Wiley-Interscience, 1984.
[54] Lutz, JT Jr. Polymeric additives for polyvinyl chloride, Owen ED., ed, Degradation and Stabilization of PVC, New York, Elsevier, Applied Science Publishers, 1984, 253–314.
[55] Nishi, T., & Kwei, TK. Improvement of the impact strength of a blend of poly (vinyl chloride) with copolyester thermoplastic elastomer by heat treatment. J Appl Polym Sci 1976, 20, 1331–1337.
[56] Sperati, CA. Physical constants of fluoropolymers, Brandrup J, Immergut EH., ed, Polymer Handbook, New York, John Wiley and Sons, 1989, V/31–V/41.
[57] James, EM. Polymer data handbook, USA, Oxford University Press, 1999.
[58] Ho, MP., Lau, KT., Wang, H., & Hui, D. Improvement on the properties of polylactic acid (PLA) using bamboo charcoal particles. Compos Part B Eng 2015, 81, 14–25.
[59] Nagarajan, V., Zhang, K., Misra, M., & Mohanty, AK. Overcoming the fundamental challenges in improving the impact strength and crystallinity of PLA biocomposites: influence of nucleating agent and mold temperature. ACS Appl Mater Interfaces 2015, 7, 11203–11214.
[60] Daitx, TS., Carli, LN., Crespo, JS., & Mauler, RS. Effects of the organic modification of different clay minerals and their application in biodegradable polymer nanocomposites of PHBV. Appl Clay Sci 2015, 115, 157–164.
[61] Burlein, GA., & Rocha, MC. LDPE/PHB blends filled with castor oil cake, AIP Conference Proceedings, AIP Publishing, 1664, 030008.

[62] Pachekoski, WM., Agnelli, JA., & Belem, LP. Thermal, mechanical and morphological properties of poly (hydroxybutyrate) and polypropylene blends after processing. Mat Res 2009, 12, 159–164.
[63] Zaman, HU., & Beg, MD. Improvement of physico-mechanical, thermomechanical, thermal and degradation properties of PCL/gelatin biocomposites: effect of gamma radiation. Radiat Phys Chem 2015, 109, 73–82.
[64] Simoes, CL., Viana, JC., & Cunha, AM. Mechanical properties of poly (ε-caprolactone) and poly (lactic acid) blends. J Appl Polym Sci 2009, 112, 345–352.
[65] Costa, MJ., Cerqueira, MA., Ruiz, HA., Fougnies, C., Richel, A., Vicente, AA., Teixeira, JA., & Aguedo, M. Use of wheat bran arabinoxylans in chitosan-based films: effect on physicochemical properties. Ind Crops Prod 2015, 66, 305–311.
[66] Srinivasa, PC., Ramesh, MN., & Tharanathan, RN. Effect of plasticizers and fatty acids on mechanical and permeability characteristics of chitosan films. Food Hydrocoll 2007, 21, 1113–1122.
[67] Li, H., Luo, Y., Qi, R., Feng, J., Zhu, J., Hong, Y., Feng, Z., & Jiang, P. Fabrication of high-viscosity biodegradable poly (butylene succinate)(PBS)/solid epoxy (SE)/carboxyl-ended polyester (CP) blends. J Appl Polym Sci 2015, 132, 42193–42199.
[68] Pivsa-Art, W., Pivsa-Art, S., Fujii, K., Nomura, K., Ishimoto, K., Aso, Y., Yamane, H., & Ohara, H. Compression molding and melt-spinning of the blends of poly (lactic acid) and poly (butylene succinate-co-adipate). J Appl Polym Sci 2015, 132, 41856–41864.
[69] Abdelwahab, MA., Taylor, S., Misra, M., & Mohanty, AK. Thermo-mechanical characterization of bioblends from polylactide and poly (butylene adipate-co-terephthalate) and lignin. Macromol Mater Eng 2015, 300, 299–311.
[70] Wibowo, AC., Misra, M., Park, HM., Drzal, LT., Schalek, R., & Mohanty, AK. Biodegradable nanocomposites from cellulose acetate: mechanical, morphological, and thermal properties. Compos Part A Appl Sci Manuf 2006, 37, 1428–1433.
[71] Torres-Rendon, JG., Schacher, FH., Ifuku, S., & Walther, A. Mechanical performance of macrofibers of cellulose and chitin nanofibrils aligned by wet-stretching: a critical comparison. Biomacromolecules 2014, 15, 2709–2717.
[72] Tang, H., Butchosa, N., & Zhou, Q. A transparent, hazy, and strong macroscopic ribbon of oriented cellulose nanofibrils bearing poly (ethylene glycol). Adv Mater 2015, 27, 2070–2076.
[73] Ismail, H., & Zaaba, NF. Tensile properties, degradation behavior, and water absorption of sago starch plastic films. J Vinyl Addit Techn 2012, 18, 235–240.
[74] Cano, AI., Cháfer, M., Chiralt, A., & González-Martínez, C. Physical and microstructural properties of biodegradable films based on pea starch and PVA. J Food Eng 2015, 167, 59–64.
[75] Wan, C., & Chen, B. Reinforcement of biodegradable poly (butylene succinate) with low loadings of graphene oxide. J Appl Polym Sci 2013, 127, 5094–5099.
[76] Broz, ME., VanderHart, DL., & Washburn, NR. Structure and mechanical properties of poly (D, L-lactic acid)/poly (ε-caprolactone) blends. Biomaterials 2003, 24, 4181–4190.
[77] López-Rodríguez, N., Lopez-Arraiza, A., Meaurio, E., & Sarasua, JR. Crystallization, morphology, and mechanical behavior of polylactide/poly (ε-caprolactone) blends. Polym Eng Sci 2006, 46, 1299–1308.
[78] Alberti, LD., Souza, OF., Bucci, DZ., & Barcellos, IO. Study on physical and mechanical properties of PHB biocomposites with rice hull ash. Mater Sci Forum 2014, 775, 557–561.
[79] Seoane, IT., Manfredi, LB., & Cyras, VP. Properties and processing relationship of polyhydroxybutyrate and cellulose biocomposites. Procedia Mater Sci 2015, 8, 807–813.
[80] Hassan, ML., Bras, J., Hassan, EA., Fadel, SM., & Dufresne, A. Polycaprolactone/modified bagasse whisker nanocomposites with improved moisture-barrier and biodegradability properties. J Appl Polym Sci 2012, 125, E10–9.

[81] Saurabh, CK., Gupta, S., Bahadur, J., Mazumder, S., Variyar, PS., & Sharma, A. Mechanical and barrier properties of guar gum based nano-composite films. Carbohydr Polym 2015, 124, 77–84.
[82] Han, SI., Lim, JS., Kim, DK., Kim, MN., & Im, SS. In situ polymerized poly (butylene succinate)/silica nanocomposites: physical properties and biodegradation. Polym Degrad Stab 2008, 93, 889–895.
[83] Gogoi, P., Horo, H., Khannam, M., & Dolui, SK. In situ synthesis of green bionanocomposites based on aqueous citric acid cured epoxidized soybean oil-carboxylic acid functionalized multiwalled carbon nanotubes. Ind Crops Prod 2015, 76, 346–354.
[84] Zhuang, W., Liu, J., Zhang, JH., Hu, BX., & Shen, J. Preparation, characterization, and properties of TiO_2/PLA nanocomposites by in situ polymerization. Polym Compos 2009, 30, 1074–1080.
[85] de Moura, MR., Lorevice, MV., Mattoso, LH., & Zucolotto, V. Highly stable, edible cellulose films incorporating chitosan nanoparticles. J Food Sci 2011, 76, N25–9.
[86] Yang, ZY., Wang, WJ., Shao, ZQ., Zhu, HD., Li, YH., & Wang, FJ. The transparency and mechanical properties of cellulose acetate nanocomposites using cellulose nanowhiskers as fillers. Cellulose 2013, 20, 159–168.
[87] Zhang, M., & Thomas, NL. Blending polylactic acid with polyhydroxybutyrate: the effect on thermal, mechanical, and biodegradation properties. Adv Polym Technol 2011, 30, 67–79.
[88] Fabra, MJ., Sánchez, G., López-Rubio, A., & Lagaron, JM. Microbiological and ageing performance of polyhydroxyalkanoate-based multilayer structures of interest in food packaging. LWT Food Sci Technol 2014, 59, 760–767.
[89] Mizuno, S., Maeda, T., Kanemura, C., & Hotta, A. Biodegradability, reprocessability, and mechanical properties of polybutylene succinate (PBS) photografted by hydrophilic or hydrophobic membranes. Polym Degrad Stab 2015, 117, 58–65.
[90] Klein, R. Laser welding of plastics: Materials, processes and industrial applications, Weinheim, Germany, John Wiley & Sons, 2012.
[91] Nampoothiri, KM., Nair, NR., & John, RP. An overview of the recent developments in polylactide (PLA) research. Bioresour Technol 2010, 101, 8493–8501.
[92] Vroman, I., & Tighzert, L. Biodegradable polymers. Materials 2009, 2, 307–344.
[93] Othman, MB., Akil, HM., Rasib, SZ., Khan, A., & Ahmad, Z. Thermal properties and kinetic investigation of chitosan-PMAA based dual-responsive hydrogels. Ind Crops Prod 2015, 66, 178–187.
[94] Avolio, R., Graziano, V., Pereira, YD., Cocca, M., Gentile, G., Errico, ME., Ambrogi, V., & Avella, M. Effect of cellulose structure and morphology on the properties of poly (butylene succinate-co-butylene adipate) biocomposites. Carbohydr Polym 2015, 133, 408–420.
[95] Nofar, M., Maani, A., Sojoudi, H., Heuzey, MC., & Carreau, PJ. Interfacial and rheological properties of PLA/PBAT and PLA/PBSA blends and their morphological stability under shear flow. J Rheol 2015, 59, 317–333.
[96] Pawar, SP., Misra, A., Bose, S., Chatterjee, K., & Mittal, V. Enzymatically degradable and flexible bio-nanocomposites derived from PHBV and PBAT blend: assessing thermal, morphological, mechanical, and biodegradation properties. Colloid Polym Sci 2015, 293, 2921–2930.
[97] Arruda, LC., Magaton, M., Bretas, RE., & Ueki, MM. Influence of chain extender on mechanical, thermal and morphological properties of blown films of PLA/PBAT blends. Polym Test 2015, 43, 27–37.
[98] de Moraes, AC., Andrade, PF., de Faria, AF., Simões, MB., Salomão, FC., Barros, EB., Do Carmo Gonçalves, M., & Alves, OL. Fabrication of transparent and ultraviolet shielding composite films based on graphene oxide and cellulose acetate. Carbohydr Polym 2015, 123, 217–227.

[99] Chen, N., Tong, Z., Yang, W., & Brennan, AB. Biocomposites with tunable properties from poly (lactic acid)-based copolymers and carboxymethyl cellulose via ionic assembly. Carbohydr Polym 2015, 128, 122–129.

[100] López, OV., Ninago, MD., Lencina, MS., García, MA., Andreucetti, NA., Ciolino, AE., & Villar, MA. Thermoplastic starch plasticized with alginate–Glycerol mixtures: melt-processing evaluation and film properties. Carbohydr Polym 2015, 126, 83–90.

[101] French, D. Organization of Starch Granules, Starch: Chemistry and Technology, Academic Press, Orlando, San Diego, New York, London, Toronto, Montreal, Sydney, Tokyo, 1984.

[102] Li, Y., & Sun, XS. Preparation and characterization of polymer– inorganic nanocomposites by in situ melt polycondensation of l-lactic acid and surface-hydroxylated MgO. Biomacromolecules 2010, 11, 1847–1855.

[103] Ray, SS., & Bousmina, M. Poly (butylene sucinate-co-adipate)/montmorillonite nanocomposites: effect of organic modifier miscibility on structure, properties, and viscoelasticity. Polymer 2005, 46, 12430–12439.

[104] Kavianinia, I., Plieger, PG., Kandile, NG., & Harding, DR. Preparation and characterization of chitosan films, crosslinked with symmetric aromatic dianhydrides to achieve enhanced thermal properties. Polym Int 2015, 64, 556–562.

[105] Figueiredo, AR., Silvestre, AJ., Neto, CP., & Freire, CS. In situ synthesis of bacterial cellulose/polycaprolactone blends for hot pressing nanocomposite films production. Carbohydr Polym 2015, 132, 400–408.

[106] Snowdon, M., Mohanty, AK., & Misra, M. Melt. Processing and characterization of bionanocomposites made from poly (butylene succinate) bioplastic and carbon black. Macromol Mater Eng 2015, 300, 118–126.

[107] Ambrosio Martín, J., Gorrasi, G., Lopez-Rubio, A., Fabra, MJ., Mas, LC., López-Manchado, MA., & Lagaron, J. M. On the use of ball milling to develop poly (3-hydroxybutyrate-co-3-hydroxyvalerate)-graphene nanocomposites (II)—Mechanical, barrier, and electrical properties. J Appl Polym Sci 2015, 132.

[108] Hou, G., Li, N., Han, H., Run, M., & Gao, J. Preparation and thermal properties of bio-based gallic acid epoxy/carbon nanotubes composites by cationic ring-opening reaction. Polym Compos 2016, 37, 3093–3102.

[109] Heinämäki, J., Halenius, A., Paavo, M., Alakurtti, S., Pitkänen, P., Pirttimaa, M., & Yliruusi, J. Suberin fatty acids isolated from outer birch bark improve moisture barrier properties of cellulose ether films intended for tablet coatings. Int J Phar 2015, 489, 91–99.

[110] Spearman, SS., Irin, F., Rivero, IV., Green, MJ., & Abidi, N. Effect of dsDNA wrapped single-walled carbon nanotubes on the thermal and mechanical properties of polycaprolactone and polyglycolide fiber blend composites. Polymer 2015, 56, 476–481.

[111] Bhuiyan, D., Jablonsky, MJ., Kolesov, I., Middleton, J., Wick, TM., & Tannenbaum, R. Novel synthesis and characterization of a collagen-based biopolymer initiated by hydroxyapatite nanoparticles. Acta Biomater 2015, 15, 181–190.

[112] Fernandes, EM., Correlo, VM., Mano, JF., & Reis, RL. Cork–Polymer biocomposites: mechanical, structural and thermal properties. Mater Des 2015, 82, 282–289.

[113] Drieskens, M., Peeters, R., Mullens, J., Franco, D., Lemstra, PJ., & Hristova-Bogaerds, DG. Structure versus properties relationship of poly (lactic acid). I. Effect of crystallinity on barrier properties. J Polym Sci B 2009, 47, 2247–2258.

[114] Risse, S., Tighzert, L., Berzin, F., & Vergnes, B. Microstructure, rheological behavior, and properties of poly (lactic acid)/poly (butylene succinate)/organoclay nanocomposites. J Appl Polym Sci 2014, 131.

[115] Camarero-Espinosa, S., Boday, DJ., Weder, C., & Foster, EJ. Cellulose nanocrystal driven crystallization of poly (d, l-lactide) and improvement of the thermomechanical properties. J Appl Polym Sci 2015, 132.
[116] Wu, W., Cao, X., Luo, J., He, G., & Zhang, Y. Morphology, thermal, and mechanical properties of poly (butylene succinate) reinforced with halloysite nanotube. Polym Compos 2014, 35, 847–855.
[117] Park, HM., Liang, X., Mohanty, AK., Misra, M., & Drzal, LT. Effect of compatibilizer on nanostructure of the biodegradable cellulose acetate/organoclay nanocomposites. Macromolecules 2004, 37, 9076–9082.
[118] Bai, H., Xiu, H., Gao, J., Deng, H., Zhang, Q., Yang, M., & Fu, Q. Tailoring impact toughness of poly (L-lactide)/poly (ε-caprolactone)(PLLA/PCL) blends by controlling crystallization of PLLA matrix. ACS Appl Mater Interfaces 2012, 4, 897–905.
[119] Xie, L., Xu, H., Chen, JB., Zhang, ZJ., Hsiao, BS., Zhong, GJ., & Li, ZM. From nanofibrillar to nanolaminar poly (butylene succinate): paving the way to robust barrier and mechanical properties for full-biodegradable poly (lactic acid) films. ACS Appl Mater Interfaces 2015, 7, 8023–8032.
[120] Zhang, B., Azuma, J., & Uyama, H. Preparation and characterization of a transparent amorphous cellulose film. RSC Adv 2015, 5, 2900–2907.
[121] Huh, M., Jung, MH., Park, YS., Kim, BJ., Kang, MS., Holden, PJ., & Yun, SI. Effect of carbon nanotube functionalization on the structure and properties of poly (3-hydroxybutyrate)/MWCNTs biocomposites. Macromol Res 2014, 22, 765–772.
[122] Murariu, M., Dechief, AL., Ramy-Ratiarison, R., Paint, Y., Raquez, JM., & Dubois, P. Recent advances in production of poly (lactic acid)(PLA) nanocomposites: a versatile method to tune crystallization properties of PLA. Nanocomposites 2015, 1, 71–82.
[123] Dou, Y., Xu, S., Liu, X., Han, J., Yan, H., Wei, M., & Duan, X. Transparent, flexible films based on layered double hydroxide/cellulose acetate with excellent oxygen barrier property. Adv Funct Mater 2014, 24, 514–521.
[124] Bower, DI. An Introduction to Polymer Physics, Cambridge University Press, Cambridge, 2002.
[125] Herrera, N., Salaberria, AM., Mathew, AP., & Oksman, K. Plasticized polylactic acid nanocomposite films with cellulose and chitin nanocrystals prepared using extrusion and compression molding with two cooling rates: effects on mechanical, thermal and optical properties. Compos Part A Appl Sci Manuf 2016, 83, 89–97.
[126] de Oliveira, RL., Da Silva Barud, H., De Salvi, DTB., Perotti, GF., Ribeiro, SJL., & Constantino, VRL. Transparent organic–Inorganic nanocomposites membranes based on carboxymethylcellulose and synthetic clay. Ind Crops Prod 2015, 69, 415–423.
[127] Ahmad, M., Hani, NM., Nirmal, NP., Fazial, FF., Mohtar, NF., & Romli, SR. Optical and thermo-mechanical properties of composite films based on fish gelatin/rice flour fabricated by casting technique. Prog Org Coat 2015, 84, 115–127.
[128] Mirzadeh, A., Ghasemi, H., Mahrous, F., & Kamal, MR. Reactive extrusion effects on rheological and mechanical properties of poly (lactic acid)/poly [(butylene succinate)-co-adipate]/epoxy chain extender blends and clay nanocomposites. J Appl Polym Sci 2015, 132.
[129] Chen, RY., Zou, W., Zhang, HC., Zhang, GZ., Yang, ZT., Jin, G., & Qu, JP. Thermal behavior, dynamic mechanical properties and rheological properties of poly (butylene succinate) composites filled with nanometer calcium carbonate. Polym Test 2015, 42, 160–167.
[130] Sarbon, NM., Badii, F., & Howell, NK. The effect of chicken skin gelatin and whey protein interactions on rheological and thermal properties. Food Hydrocoll 2015, *45*, 83–92.
[131] Chanvrier, H., Chaunier, L., Della Valle, G., & Lourdin, D. Flow and foam properties of extruded maize flour and its biopolymer blends expanded by microwave. Food Res Int 2015, 76, 567–575.

[132] Plackett D Ed., Biopolymers: new materials for sustainable films and coatings, West Sussex, UK, John Wiley & Sons, 2011.
[133] Huang, SJ., & Edelman, PG. An overview of biodegradable polymers and biodegradation of polymers, Degradable Polymers, Springer, Dordrecht, 1995, 18–28.
[134] Premraj, R., & Doble, M. Biodegradation of polymers. Indian J Biotechnol 2005, 4, 186–193.
[135] Siracusa, V., Rocculi, P., Romani, S., & Dalla Rosa, M. Biodegradable polymers for food packaging: a review. Trends Food Sci Technol 2008, 19, 634–643.
[136] Itävaara, M., & Vikman, M. An overview of methods for biodegradability testing of biopolymers and packaging materials. J Environ Polym Degrad 1996, 4, 29–36.
[137] Grima, S., Bellon-Maurel, V., Feuilloley, P., & Silvestre, F. Aerobic biodegradation of polymers in solid-state conditions: a review of environmental and physicochemical parameter settings in laboratory simulations. J Polym Environ 2000, 8, 183–195.
[138] Manay, NSO. Food: facts and principles, New Age International publishers, New age international limited publishers, Darya Ganj, New Delhi, 2001.
[139] Chouliara, E., Karatapanis, A., Savvaidis, IN., & Kontominas, MG. Combined effect of oregano essential oil and modified atmosphere packaging on shelf-life extension of fresh chicken breast meat, stored at 4 C. Food Microbiol 2007, 24, 607–617.
[140] Hanes, M. J. inventors, Standard Oil Company, assignee, Vol. US*4260311*, 1981.
[141] McMillin, KW. Where is MAP going? A review and future potential of modified atmosphere packaging for meat. Meat Sci 2008, 80, 43–65.
[142] Brody, AL. Meat Packaging: Past, Present and Future. Presentation at 55th reciprocal meat conference, East Lansing, Michigan, USA, 2002.
[143] Nerín, C., Tovar, L., Djenane, D., Camo, J., Salafranca, J., Beltrán, JA., & Roncalés, P. Stabilization of beef meat by a new active packaging containing natural antioxidants. J Agric Food Chem 2006, 54, 7840–7846.
[144] Park, SI., Marsh, KS., & Dawson, P. Application of chitosan-incorporated LDPE film to sliced fresh red meats for shelf life extension. Meat Sci 2010, 85, 493–499.
[145] Scannell, AG., Hill, C., Ross, RP., Marx, S., Hartmeier, W., & Arendt, EK. Development of bioactive food packaging materials using immobilised bacteriocins Lacticin 3147 and Nisaplin®. Int J Food Microbiol 2000, 60, 241–249.
[146] Mohamed, S., Taufik, B., & Karim, MNA. Effects of modified atmosphere packaging on the physicochemical characteristics of ciku (Achras sapotaL) at various storage temperatures. J Sci Food Agric 1996, 70, 231–240.
[147] Ambaw, A., Beaudry, R., Bulens, I., Delele, MA., Ho, QT., Schenk, A., & Verboven, P. Modeling the diffusion–Adsorption kinetics of 1-methylcyclopropene (1-MCP) in apple fruit and non-target materials in storage rooms. J Food Eng 2011, 102, 257–265.
[148] Chauhan, AK., & Patil, V. Effect of packaging material on storage ability of mango milk powder and the quality of reconstituted mango milk drink. Powder Technol 2013, *239*, 86–93.
[149] Thakur, BR., & Arya, SS. Relative suitability of plastic films for the frozen storage of mango pulp. J Food Process Pres 1988, 12, 171–178.
[150] Romphophak, T., Siriphanich, J., Promdang, S., & Ueda, Y. Effect of modified atmosphere storage on the shelf life of banana 'Sucrier'. J Hortic Sci Biotechnol 2004, 79, 659–663.
[151] Ducruet, V., Fournier, N., Saillard, P., Feigenbaum, A., & Guichard, E. Influence of packaging on the aroma stability of strawberry syrup during shelf life. J Agric Food Chem 2001, 49, 2290–2297.
[152] Jacomino, AP., Kluge, RA., Sarantópoulos, CIGDL., & Sigrist, JMM. Evaluation of plastic packages for guava refrigerated preservation. Packag Technol Sci 2001, 14, 11–19.

[153] Dash, KK., Goswami, TK., & Deka, SC. Modeling for the modified atmosphere packaging of Sapota fruit. Agri Eng Int: CIGR J 2012, 14, 102–109.
[154] Almenar, E., Samsudin, H., Auras, R., Harte, B., & Rubino, M. Postharvest shelf life extension of blueberries using a biodegradable package. Food Chem 2008, 110, 120–127.
[155] Benyathiar, P., Almenar, E., Auras, R., Ryser, E., Harte, J., & Siddiq, M. A sustainable package for fresh cut green salad, United Fresh 2009 Convention, Las Vegas, Nevada, *US*, 2009.
[156] Ghaouth, A., Arul, J., Ponnampalam, R., & Boulet, M. Use of chitosan coating to reduce water loss and maintain quality of cucumber and bell pepper fruits. J Food Process Preserv 1991, 15, 359–368.
[157] Campaniello, D., Bevilacqua, A., Sinigaglia, M., & Corbo, MR. Chitosan: antimicrobial activity and potential applications for preserving minimally processed strawberries. Food Microbiol 2008, 25, 992–1000.
[158] Zhang, D., & Quantick, PC. Effects of chitosan coating on enzymatic browning and decay during postharvest storage of litchi (Litchi chinensis Sonn.) fruit. Postharvest Biol Technol 1997, 12, 195–202.

Narendren Soundararajan, Shasanka Sekhar Borkotoky
and Vimal Katiyar

7 Up-to-date advances of biobased and biodegradable polymers in food packaging

7.1 Introduction

In this modern era, plastic packaging is a pervasive and essential part of product supply chain. It preserves, protects and even enhances the quality of the product during its transfer from processing to storage till it is utilized or consumed by the end user. Packaging is around 2% of the gross national product of developed countries, where 50% is used to pack food-based products [1]. Food packaging is the soul of modernized food industries as package performs a sequence of tasks; primary of them worth mentioning are containment, safety, handiness and communicating information. These functions must be assessed and considered simultaneously during package development process as all are interconnected. The packaging has to perform its functions in three different environments: (1) physical environment where the packaging safeguards the food items from any physical damage, (2) ambient environment denotes the oxygen, water vapor, light, microorganism and macroorganisms and (3) human environment in which the package interacts with people. A typical example can be carton of milk that consists of paperboard lined with plastics that acts as an effective barrier against the outside environment. Later, the carton communicates with the consumer via printed branding and nutritional information.

Innovations in the field of food packaging involves contribution of multitude of professionals from engineering, microbiology, food science, chemistry, design and others. New technologies in food packaging are emerging in response to consumer demands or trends of industries toward increasing the shelf life with controlled quality of mildly preserved, fresh, tasty and convenient food products. In addition to this, change in the consumer lifestyle, retailing practice that acts as major challenge to food packaging industries, also proves to be a driving force for the development of novel and better ideas in packaging sector to increase the shelf life [1]. In recent years, the terms such as "biodegradable," "biocompatible," "compostable," "environmental friendly," "annually renewable," "sustainable," "green" and "biopolymers" have become some commonly used keywords in literatures related to food safety and food packaging, which depict the increasing concerns about the nonrenewable plastic wastes and end-of-life disposal problem worldwide. Exception to paper and its related goods, traditional packaging material in food industry relies on noncompostable or fossil-based plastics. This tradition started to change at the beginning of the twentieth century, where packaging-related materials such as inks, dyes,

https://doi.org/10.1515/9783110648034-007

paints, pharma products, chemicals and plastics were derived from biological origin. Products derived from plants and by-products of fermentation are seriously researched to produce sustainable biobased products.

An increase in municipal solid waste (MSW) from 179.6 million tons in 1988 to 245.7 million tons in 2005 was due to use of nonrenewable resources, of which the total solid waste of 31.2% was from packaging [2]. The current methods of plastic solid waste disposal are dominated by landfilling, incineration and composting and some of them are recycled. Still there is a long way to minimize the amount of plastic materials existing in MSW. The use of biopolymers is found to be an opportunistic way to develop eco-friendly and biodegradable packaging materials for food [3].

In a year, only 3.5% (6 billion tones) of biomass are utilized by mankind from the total production of about 170 billion tones. This 6 billion tones include majorly food, while rest are used for energy, paper, furniture and construction and the remaining 5% are for chemicals and clothing [4]. There are numerous sources of biopolymers available such as starch cellulose, chitosan and various proteins from plant and animal. They are considered to be an attractive alternative for nonrenewable fossil fuel-derived plastic packaging material because of its availability, renewability, low cost, eco-friendliness, bio-degradability and compostability. Like petroleum-derived packaging materials, biopolymers also show promising barrier properties to gas and water vapor and thereby maintain food quality and extend shelf life [4]. Further, biopolymer films seem to be an outstanding matrix to incorporate wide variety of fillers to enhance antioxidants, antifungal, antimicrobial, color and nutritional value of the plastics [4].

Biopolymers are advantageous over its counterpart as they can degrade under controlled environment. However, relatively low mechanical strength and high moisture sensitivity limit the usage of biopolymer films in commercial front [5]. Although protein and carbohydrate films exhibit good barrier properties against oxygen in certain environmental condition and are also found to provide better mechanical strength, their hydrophilic behavior show least barrier against water vapor. Several efforts that focused on property improvement of biopolymer-based films pose no satisfactory results toward physical, thermal and mechanical properties of these films, which pose a challenge for its industrial applications.

Researchers experimented on techniques like chemical modification, addition of plasticizer, blending two biopolymers for improved characteristics and increased miscibility of two or more incompatible polymers by incorporating compatibilizers to decrease interfacial energy and stabilizing polymer blends for improved biopolymer properties. Bionanocomposites that are one step ahead of existing material show comparable mechanical properties, barrier as well as thermal properties. Bionanocomposite refers to the reinforcement of biopolymer matrix with nanoparticles as fillers with one of its dimensions in the nanometer range (1–100 nm). Because of its high aspect ratio and large surface area, these bionanocomposites show improvement in properties when compared with pristine polymers or conventional micro- or macro-scale composites [5].

7.2 Natural biopolymers for food packaging

7.2.1 Starch-based packaging materials

Starch is easily available and is a low-cost biopolymer used in food-packaging applications. Plants such as corn, cereal grain and so on store major energy in the form of starch. Starch-based biodegradable films are used as edible films and coatings. Starch sources like high amylose starch and modified starch [6] can form self-standing films from aqueous solution, and show reasonable gas barrier properties but least water barrier and poor mechanical strength, thus limiting its applications. Starch has been used as reinforcement material in plastics [6]. The addition of small amount of starch (6–30%) into various biopolymers increases the overall biodegradability of the product. It can behave as thermoplastic on addition of plasticizer like water and glycerol. Thermoplastic starch (TPS) products are produced by varying viscosity of the material, solubility and water adsorption capacity and by varying the composition of moisture or plasticizer content or by changing the ratio of amylose/amylopectin in the raw material and by changing the temperature and pressure in the extruder. Further, extensive research has been carried out on TPS plasticization along with glycerol, sorbitol, urea, formamide and dimethyl sulfoxide, and low molecular weight sugar has been researched with great deal for the processability of TPS. They are mainly used as compostable foams that are soluble, thermally expanded trays, molded parts and expanded layers as an alternative for polystyrene (PS) foams. BIOTEC (Emmerich, Germany) has product lines for TPS that includes Bioplast®, Bioflex® and Biopur® [6].

7.2.2 Cellulose-based packaging materials

Cellulose, the most widely available polymer on the Earth, is essentially a linear polymer of anhydrous glucose. Woods provide the major source of polymer, which is about 40–50% by weight. In spite of its abundance on Earth, it is not suitable for the production of films because of its high crystallinity, fibrous nature and insolubility in water. Cellophane is a product of cellulose produced by dissolving cellulose, sodium hydroxide and carbon disulfide added into sulfuric acid to make films. Cellophane shows a great degree of mechanical property, but its sensitivity to moisture is the major limitation toward its widespread application as packaging material. To improve its moisture barrier properties, cellophanes are frequently coated with nitrocellulose wax or polyvinylidene chloride. These coated cellophanes are mostly used for packing freshly baked goods, processed meat, cheese slice and candy. Due to its nonthermoplastic nature, it is not heat sealable [6]. Cellulose acetate, ethyl cellulose, hydroxylethyl cellulose and hydroxyl-propyl cellulose are commercially available derivatives of cellulose. Even though these derivatives are not optimal for food packaging, cellulose acetate films allow respiration and reduces fogging [6]. Mazzucchelli (Castiglione Olona, Italy) and Planet Polymer (CA, USA) manufacture biodegradable plastics products from cellulose acetate under the trade names of BIOCETA® and Enviro Plastic®, respectively. BIOCETA® is also involved in the manufacturing of biodegradable packaging films, stretchable films and tubes [6].

7.2.3 Chitosan- and pectin-based packaging materials

Chitosan ranks second to cellulose in terms of abundancy. Chitosan is also known to be soluble chitin, a product obtained by deacetylation of chitin. The repeat unit 1,4 linked 2-deoxy-2-aminoglucose forms the chitosan. Chitosan films formed by solution casting are clear, tough and flexible and provide good oxygen barrier. Chitosan films can act as antifungal element and prevent food from spoilage and can change atmosphere for freshly picked fruits. In order to improve the mechanical strength and barrier properties of the chitosan films, researchers studied the effect of factors like concentration of plasticizer, time of storage, type of acid and its concentrations used for deacetylation, change in molecular weights of chitosan and the degree of deacetylation [6].

Pectin, a polysaccharide, belongs to a family of heterogeneous branched polymers that mainly consist of methylated galacturonan segments supported by rhamnose residues that are linked to short neutral sugar side chains. Pectin can be modified readily by demethylation and is used for the preparation of edible film in the presence of calcium ions [6]. However, pectin films are not widely used in food packaging as it shows high water vapor permeability and hence is usually combined with other biopolymers for application [6]. Pectin coated on food is used as packaging material and shows no adequate moisture barrier properties, but they delay water loss from the food product without dehydrating the food significantly. Further investigation shows that pectin coating has the ability to retard lipid migration and to improve appearance and handling of food products.

7.2.4 Polyhydroxyalkanoate (PHA)-based packaging materials

Under limiting nutritional conditions, various microorganisms accumulate PHAs, a polymer of hydroxyalkanoates by utilizing carbon source. There exist around 300 different microorganisms that are capable of synthesizing and accumulating PHAs [7].

These synthesized biopolymers mainly include polyhydroxy-butyrates and copolymers of hydroxyl-butyrate (HB) and hydroxyl-valerate (HV) that are commercially produced by Metabolix, under the trade name Biopol^{TM}. Wella AG (Darmstadt, Germany) is manufacturing blow-molded PHA bottles using Biopol for shampoo packaging [7]. Nodax trademarked denotes polyhydroxybutyrate (PHB) co-hydroxyalkanoates developed by Procter and Gamble. PHA has exceptional film-forming properties and can easily form a coat that is comparable to the properties of polypropylene (PP) [7]. By changing the ratio of HB and HV, the properties of the film can be adjusted. At high concentrations, PHB together with polyhydroxyvalerate (PHV) form strong and flexible films. Citrate ester in poly(3-hydroxybutyrate-*co*-hydroxyvalerate) (PHBV) can give plasticization effect [7]. As PHAs are hydrophobic in nature, they show better moisture barrier properties. Even though PHAs are biodegradable under soil condition and holds outstanding processability, they face difficulties in reaching market because of its high cost and poor gas barrier properties [7].

7.2.5 Polylactic acid (PLA)-based packaging materials

PLA shows high potential for commercialization among all other polyesters that have biological origin. Carothers in 1932 manufactured PLA by ring-opening polymerization, which was further improved by Dupont and Ethicon [8] to produce PLA at large scale. PLA, a linear aliphatic polyester, is produced from monomers of lactic acid. The ester linkages are more prone to cleavage by enzyme hydrolysis or chemical hydrolysis. Fermentation of glucose from various sources like starch from corn or wheat or lactose from whey or sucrose from molasses can be utilized to produce lactic acid at low cost. Cargill Dow offers a variety of PLA products (Nature Works®) that are manufactured from renewable sources like corn starch or sugar beets. Mitsui Chemicals (Japan) and Shimadzu (Japan) are mentionable PLA producers. PLA shows good mechanical and thermal properties; these properties highly depend on ratio between meso-forms D- and L-PLA. PLA is known to show plasticizing effect on the addition of polyethylene glycol, triethyl citrate and partial fatty acid esters [9, 10]. PLA exhibits moderate water and oxygen permeability values similar to PS. Due to its high density, it delivers poor resistance to heat, high density, highly polarity and brittleness. It is because of this limiting factors that commercialization is a problem.

The PLA from Nature Works is extrusion grade, which can also be thermoformed or blow molded into useful products similar to commercial non-renewable polymers. The film-grade PLA can be used in candy wrapping, lamination packing, packing pet foods and so on. The thermoforming grade of PLA can be used for a number of applications such as dairy cups, food service wares, blister packs and so on.

7.3 Case studies: up-to-date advances in biodegradable food packaging materials

7.3.1 Bionanocomposites in food packaging

Several researchers are actively involved in analyzing the performance of biopolymers by adding several inorganic nanoparticles as additives. Various nanofillers currently being developed are nanoclay (layered silicates) [11, 12], cellulose nanowhiskers [13–15] and carbon nanotubes [16]. Among these, the layered inorganic silicates like clay has received much attention in the plastic manufacturing industry. This is not only due to their environmental friendliness, natural abundance and low cost but also due to their significant enhancements and relatively simple processability. Due to the sustainability of the nanocomposites, they have been extensively studied for food-packaging application that includes production of beverage and beer bottles by injection molding, cartons coatings for juice packaging, cast and blown films [17–19]. Incorporation of nanocomposites in food packaging shows a promising increase in oxygen-scavenging properties, high resistance against heat and improved oxygen barrier properties that can increase food shelf life from 3 to 5 years [20, 21].

7.3.2 Properties of PLA bionanocomposites

The dispersion of fillers in the polymer matrix can play an important role in the alteration of polymer matrix properties. On adding clay which is not modified and organically modified clays (C20A and C30B) in the PLA polymeric matrix, a decline in tensile strength and elongation with 5% clay loading was observed [22]. This decline in tensile modulus could be due to the poor dispersion of the fillers within polymer matrix. A slight increase in the properties of the polymer composite was observed on addition of organoclay to the matrix. The aspect ratio of the filler material added to the polymer matrix can be the reason for such improvement in properties.

PLA/cellulose nanocrystals (CNC) nanocomposite also resulted in a similar trend in tensile properties compared to layered silicate nanocomposites. The rigidity and high surface area of the nanostructure attributed to this increased elastic modulus and elasticity [23]. Surprising results were obtained when PLA/acetylated CNC (ACNC) nanocomposites resulted in an increase in elastic modulus but the fraction strain decreased. A 6 wt% loading of fillers increased the tensile strength at break. At higher concentration of the nanofillers, they tend to form aggregation, thus resulting in low values of strength and elongation of the nanocomposite. The addition of nanofiller to the PLA polymer matrix makes a tortuous pathway in the composite film that can improve barrier properties. The sheet-like morphology layered-silicates and cellulose nanowhiskers are predominantly efficient because of their high aspect ratio in length to width [23, 24]

Four different nanocomposites of PLA as matrix were prepared and a comparison in their barrier properties was done for better understanding [25]. In this case, four kinds of organically modified clays montmorillonites, saponite and mica with different types of interlayer cations were used. A 5–10 wt% loading of organoclays to the PLA matrix by traditional methods showed around 60% reduction in oxygen permeability. On increase of the loading above 10 wt%, the clay particles tend to aggregate, which caused a decrease in properties. By changing the nature of the nanofiller mixed to the PLA and with a modified mixing procedure, the ductile nature of the nanofiller can be overcome by gaining specific clay location in the PLA matrix [26, 27]. PLA/natural rubber (NR) blend is studied for O_2 and CO_2 permeability on adding desired concentration of organoclays C15A and C30B. There is a steep reduction in the value of gas flux from P O_2 = 1.01 of PLA/NR to P O_2 = 0.86 in PLA/NR/C30B 3 wt% composite. It was clearly observed that the inclusion of C30B clay to the blend improved the gas barrier properties tremendously, which was because of its better compatibility. The gas barrier properties of PLA/CNC could be compared with PLA/cellulose microfibers (MFC) by keeping the concentration constant [26].

Bilayer and trilayer filler systems of composite are prepared with PLA and they showed a decrease in oxygen permeability [27, 28]. In the bilayer system, filtered 2,2,6,6-tetramethylpiperidinyloxy radical oxidized with CNF shows a reduction in permeability value from 746 $mL/m^2/day^1\ Pa^{-1}$ to 1 $mL/m^2/day^1$ due to the presence of oxidized CNF thin layer. In the trilayer systems glycerol and plasticized gelatin are used as in inner layer and showed a reduction of 17.1 ± 2.3 $cm^3.mm/m^2/day^1$ oxygen permeability compared to neat PLA. The layer-by-layer and multilayer approaches of nanocomposite production seem to be latest and show promising properties [22, 29, 30]. Clay and chitosan were attracted to the PLA film surface by electrostatic forces and a multilayer composite is formed, which shows about 97% decrease in oxygen permeability. The thickness of the layer and decrease in permeability are proportional. The addition of silylated CNC into PLA matrix causes a great increase in the nuclei density, which is due to the high crystallinity and uniform dispersion of the CNC in the PLLA matrix [22]. In addition, interference is observed as a result of an increase in nucleating particles. PLA with the addition of layered nanosilicate shows a single-step weight loss, with maximum decomposition value at 370 ℃. The layered nanosilicate acted as a barrier for the gases that are formed during the degradation process, thus improving the thermal stability of PLA [31, 32].

7.4 Biodegradable polymeric foams

Foams can be defined as the class of material that contains gaseous voids present in it. The origin of the word "foam" comes from medieval German word "Veim," which represents froth [33]. Foams possess some uniqueness, which clearly differentiates them from its nonfoamed counterparts. Foams can be fabricated from all three types of materials: metals, ceramics and elastomers. Elastomeric foams have a wide area of applications compared to the other two classes at present time. Different polymeric materials can be foamed to convert them to lower density materials for some useful applications such as insulation, convenience and comfort, weight reduction and so on. Polymeric foams can be classified in two classes: thermoplastic foams and thermoset foams. Thermoplastic foams can be recycled and also they can have the ability to be reprocessed; however, thermoset foams cannot be recycled or reformed. Foam generally consists of porous structure that comprises voids (gaseous) surrounded by dense phase or a composite of gas/solid. A typical field emission scanning electron micrograph (FESEM) image of PLA foam surface is shown in Figure 7.1. The special properties of foam are due to the difference in nature between solid and gas [34]. The three unique characteristics of foam are: low density, light weight and less material usage, which are the main advantages of foam in terms of application purposes. Two different terminologies have been used in foaming industries: we should not confuse with the terms foam and foaming, the first one is the material and the later one is the technology to fabricate foams.

7.5 Development of polymeric foams

The elastomeric foams have the major areas of utilization in the day-to-day life. They have wide applications in the society from thermal/acoustic insulations and applications like cushioning to various advanced applications such as tissue engineering. Natural foams were known for a long time, but the production of synthetic polymeric foams was introduced in the market almost 60 years ago. The development of polymeric foams is now a major area of research. Generally, nowadays, whenever new polymers are developed, they are quickly foamed to investigate the useful properties in its expanded foamed state. Various properties can be tuned to meet particular requirements and can increase substantially the potential applications of a polymer in biomedical and commodity applications. Some principles and concepts can be applied in all the foams universally. Some established methods for polymeric foam preparation include extrusion, molded bead, cross-linked foam, injection molding, reactive foaming and gelation. In current days, polymeric foam is a well-developed and well-established branch in polymeric industry as a whole.

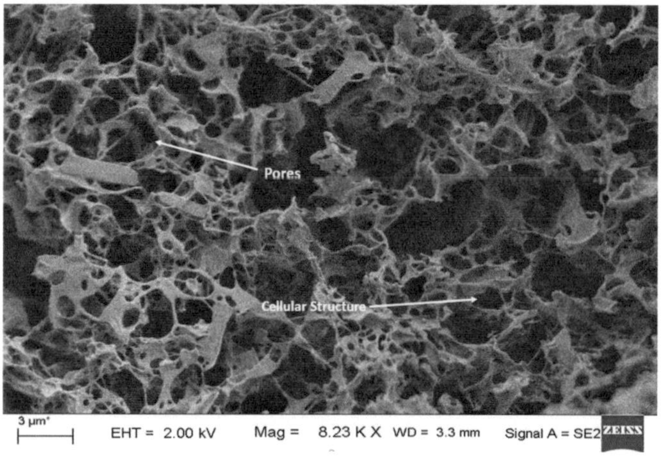

Figure 7.1: FESEM image of PLA-based foam showing the pores and its cellular morphology fabricated in Centre of Excellence for Sustainable Polymers (CoE-SusPol), IIT Guwahati.

Polymeric foams can be seen from different perspectives like product, technology and components. Polymeric foam products can be classified on the basis of density, cell size, dimension, cell density, morphology and property [35]. If we go to the history of development of elastomeric foams, we find that from 1930s to 1950s, a strong foundation was made in terms of technology and principles for foam industry. The peak development of the foam industry happened in 1960s through the 1980s. Just within 25 years, the foam industry developed from lab scale to pilot plant and finally to commercial success. The first foam developed was PS foam (1931), which was followed by polyurethane foams in 1937. After that lots of developments took place in foam industry. However, the first biobased polymeric foam was developed in 1964; it was starch foam. The latest addition in biodegradable foams is PLA foam, which is a promising material for foaming in various applications [36].

7.5.1 Some potential biobased polymers for foaming

PLA is one of the popular biodegradable polymers. It is already in use as a packaging material. However, its foamed counterparts can also be used as a packaging material. PLA is already used as a packaging material for various containers, for drinking cups, lamination films and blister packages. PLA is basically developed for antimicrobial food packaging [37].

PHA is one of the polymers that can be used in packaging and coating purposes. Various useful commodity products can be fabricated from PHA such as shopping bags, cosmetic containers, medical surgical garments, compostable bags, cups, paper coatings and so on.

PHB has the potential to be used as a packaging material in near future. Recent studies show that PHB can replace PP at high-temperature applications. However, in normal refrigeration and freezing condition, PP is better than PHB. Polycaprolactone, amorphous PLA, PHBV and some of their nanocomposites can also be used for food packaging applications.

7.6 Sustainable food packaging using biobased polymeric foams

Biobased polymeric foams can be a potential candidate for sustainable food packaging. The main motive of sustainable packaging is to replace nonrenewable resources to renewable ones. It is generally observed that foodstuffs are damaged due to factors like chemical, physical and microbiological spoilage, which lead to loss of food qualities and affect their color and also the flavor. Plastic materials have a major contribution to packaging sector. Due to the emerging environmental issues, food packaging demands sustainable packaging, which will finally reduce energy and material usage. It will also contribute to the society by reducing greenhouse gases by using renewable source materials for packaging. Biodegradable PLA foams also have a comparable mechanical property with the conventional foams like PS. They have good barrier properties and are heat-sealable to prevent food from contaminations. Other added advantages include easy to process and are available in different forms (e.g., film, sheet and foam), which will help them to be applicable in the wide fields of food packaging.

7.7 Processing technology for foam fabrication

Polymer foaming involves mainly three steps:
- bubble formation in the liquid system;
- bubble growth and
- bubble stabilization.

Foaming process is mainly done by two processes:
- soluble/physical foaming and
- reactive foaming.

7.7.1 Physical/soluble foaming

This process involves the dissolution of physical blowing agents (PBA) in the polymer melt. Common example of this process is the extrusion and injection molding with PBA. This process is capable of producing high-speed continuous processing of foams.

7.7.2 Casting and leaching

This technique involves dissolution of the polymer into a highly volatile solvent and thereafter casting the prepared solution into a mold (of different shapes of requirements) that contains a solid porogen. The porogen is particle that is a water-soluble salt, for example, NaCl, KCl, etc. which can be washed out after the solvent has been evaporated. This will ultimately leave a porous and foamed polymeric structure, which contains gaseous voids in its structure. The pore size and morphology of the fabricated foam can be changed in this method by controlling the size and distribution of the porogen particles and the amount added in the solution. The main disadvantage of this process is contamination and the use of solvents [38].

7.7.3 Foaming using gases

In this technique, PBAs are used for foaming the polymer. Problems concerning the environment and the need for developing environmental friendly foaming agents have triggered research in this area. Here mainly inert gases like nitrogen and carbon dioxide are used as a blowing agent, where carbon dioxide acts as a greener solvent. The factors effecting the selection of PBAs include following:
- Physical: volatility, critical temperature, latent heat and so on
- Chemical: reactivity and stability
- Transportation: permeability and diffusivity
- Safety: toxicity and flammability
- Environmental issues
- Availability: cost and storage

The main advantages of foaming by using gases technique are that it is a nontoxic and environmental friendly process. Although nitrogen is also used as a blowing agent, a greater use of carbon dioxide as a foaming agent has been found as CO_2 helps in enhancing the processability of the polymer. The use of carbon dioxide in supercritical stage has some added advantages. Because in supercritical fluid there is no distinct liquid and gas phases, it can effuse through solids like a gas, and can dissolve materials like a liquid.

7.7.4 Thermally induced phase separation (TIPS)

TIPS process mainly involves two steps. Polymer pellets with blowing agents are partly foamed in steam in the first step. These foamed pellets are then transferred into a mold and again come into contact with steam, in the second step, which results in extra foaming of the pellets. The pellets stick together and take the shape of mold due to the further expansion. Generally, a low boiling liquid such as pentane and hydro chlorofluorocarbons is used as a foaming agent in this technique.

7.7.5 Reactive foaming

In this technique, the final foam production is yielded from a chemical reaction. It initially require a certain amount of gases, which will finally form a foamed structure. Chemical-blowing agents (CBAs) are used in this method. CBAs are chemicals that are capable of liberating gaseous components via reactions and/or thermally induced decomposition [39].

Considerations of CBAs are as follows:
- Processing capability with base polymer
- Type and amount of decomposed gas
- Appropriate decomposition with other additives
- Color from leftover or by-products
- Nucleating effects out of decomposed particles

Examples include azodicarbonamide, titanium hydride and isocyanates.

7.8 Foam packaging

Foam packaging is basically used on shock absorber theory, protecting a moving packaged object from damage. It is normally used for delicate electronic equipment. The physical performance of foams can be fine tuned to make it useful for packaging purposes. The market is mainly dominated by PS foams due to its various favorable properties. Flexible foams can be easily cut and shaped for bracing, supporting and wrapping objects. Conventional polymeric foams like polyurethane foams are basically used for packaging purposes. EarthShell Corp. commercially produces food laminate products, which are primarily used for serving foodservice industry with food containers and serviceware. In general, some foams were classified on the basis of their ability to absorb shock [40]. Polyurethane foam is mainly utilized as packaging material for electronic parts sensitive to static discharge. For the above-mentioned purpose, PU foams are modified by using antistatic chemical agents. Various foam properties like cell size and cell density play a major role in the area of use of that particular foam. It may affect the cushioning that can absorb shock or impact dissipation particularly used for fragile object. The main advantages of foam material for packaging are their low density and light weight, and this will allow them to be used in buoyancy applications, such as aircraft, space vehicles, skis, racing yachts and portable buildings.

7.9 Developments in sustainable foams

The main bioplastic foams used for packaging include starch, PLA, poly(3-hydroxybutyrate-*co*-3-hydroxyvalerate), poly(hydroxyesterether) and polyhydroxyl alkanoates. However, the combination of the above polymeric materials is also to be studied for foam-packaging applications.

For food packaging applications, Green Cell Foams® (GCF) are now available in the market for commercial purposes. Biodegradable GCF is made from a proprietary corn-starch blend. Samina et al. recently investigated biodegradable GCF and targeted it for protective applications. They noticed that dimensional stability is the crucial factor for protective foam packaging-applications [41].

Starch foam is the most widely used biofoam till now for cushioning and packaging applications. Lots of research have been done to improve its various mechanical properties (impact, compression, tension and flexural behavior) so that it can be utilized in a wide area of applications. It was observed that moisture levels and extrusion temperature affect properties of starch foams. Also the blending of starch with various other bioplastics was investigated. It was observed that the incorporation of PLA, PHEE and PHBV promoted higher expansion ratio and lower density foams. The foam prepared was found to be more water resistant than normal starch foams [42]. Another interesting investigations on starch/PLA hybrid foams was done by Preechawong et al. They have mainly studied the effects of relative humidity, storage time, PLA content and type and content of added plasticizer (e.g., glycerol, urea or ammonium chloride) on moisture and water absorption, mechanical properties and enzymatic degradability of the prepared foams. They fabricated the hybrid foams by baking a mixture of starch, PLA and other ingredients in a hot mold. They used compression molding technique to prepare the foams [43]. Lee et al. investigated the preparation and characterization of tapioca starch–PLA nanocomposite foams. They used melt intercalation technique. They used four different types of organoclays: Cloisite 10A, Cloisite 25A, Cloisite 93A and Cloisite 15A to produce nanocomposite foams. It was observed that Young's modulus (E) of the tapioca starch–PLA nanocomposite foams was significantly affected ($p < 0.05$) by the type of organoclay [44].

Another interesting investigation on starch foam was done by Guan et al. on extrusion foaming of PLA and acetylated starches blend. They used 5 wt% talc and different amounts of ethanol. They observed that increasing PLA concentration led to more foam expansion, which is suitable for some packaging applications. It was also found that ethanol acted both as a blowing agent and also as a solubilizing agent to homogenize PLA and starch [45].

Another bioplastic, PLA foam is now showing a promising impact in the foam industry due to its various properties comparable with nondegradable counterparts. Lot of literatures are available for PLA foams in various packaging applications. Ajioka et al. explored PLA foams for cups, thermal insulators, disposable food trays and cushioning materials [46]. Researchers also investigated on PLA/PS blend foam and then created an interconnected porous structure by first foaming the PLA/PS blend and then extracting the PS phase. PS phase extraction ensured the inter pore connectivity of the cells. The fabricated foam has tissue-engineering applications [47]. Investigations were also done in expanded PP/PLA blend bead foams. Tang et al. investigated an autoclave batch-foaming process for the fabrication of expanded PP/PLA expanded bead foams to get nice ellipse shaped, well-defined cell structure and a high cell density for various packaging applications [48].

References

[1] Andersson, SR., Hakkarainen, M., Inkinen, S., Södergård, A., & Albertsson, AC. Polylactide stereocomplexation leads to higher hydrolytic stability but more acidic hydrolysis product pattern. Biomacromolecules 2010, 11, 1067–1073.

[2] Marsh, K., & Bugusu, B. Food packaging-roles, materials, and environmental issues. J Food Sci 2007, 72, R39–55.

[3] Jamshidian, M., Tehrany, EA., Imran, M., Jacquot, M., & Desobry, S. Poly-Lactic Acid: Production, applications, nanocomposites, and release studies. Compr Rev Food Sci Food Saf 2010, 9, 552–571.

[4] Robertson, GL. Food packaging: principles and practice, CRC press, Boca Raton, Florida, US, 2005.

[5] Han JH ed, Innovations in food packaging, Elsevier Academic Press, San Diego, California, USA, 2005.

[6] Tang, XZ., Kumar, P., Alavi, S., & Sandeep, KP. Recent advances in biopolymers and biopolymer-based nanocomposites for food packaging materials. Crit Rev Food Sci Nutr 2012, 52, 426–442.

[7] Alavi S, Thomas S, Sandeep KP, Kalarikkal N, Varghese J, Yaragalla S eds, Polymers for packaging applications, CRC Press, 2014.

[8] Avella, M., De Vlieger, JJ., Errico, ME., Fischer, S., Vacca, P., & Volpe, MG. Biodegradable starch/clay nanocomposite films for food packaging applications. Food Chem 2005, 93, 467–474.

[9] Bitinis, N., Verdejo, R., Maya, EM., Espuche, E., Cassagnau, P., & Lopez-Manchado, MA. Physicochemical properties of organoclay filled polylactic acid/natural rubber blend bionanocomposites. Compos Sci Technol 2012, 72, 305–313.

[10] Busolo, MA., Fernandez, P., Ocio, MJ., & Lagaron, JM. Novel silver-based nanoclay as an antimicrobial in polylactic acid food packaging coatings. Food Addit Contam 2010, 27, 1617–1626.

[11] Cabedo, L., Luis Feijoo, J., Pilar Villanueva, M., Lagarón, JM., & Giménez, E. Optimization of biodegradable nanocomposites based on aPLA/PCL blends for food packaging applications, Macromolecular Symposia, Wiley-VCH Verlag, Weinheim, 2006, Vol. 233, 191–197.

[12] Choudalakis, G., & Gotsis, AD. Permeability of polymer/clay nanocomposites: A review. Eur Polym J 2009, 45, 967–984.

[13] Conn, RE., Kolstad, JJ., Borzelleca, JF., Dixler, DS., Filer, Jr LJ., LaDu, Jr BN., & Pariza, MW. Safety assessment of polylactide (PLA) for use as a food-contact polymer. Food Chem Toxicol 1995, 33, 273–283.

[14] Coulier, L., Kaal, ER., Tienstra, M., & Hankemeier, T. Identification and quantification of (polymeric) hindered-amine light stabilizers in polymers using pyrolysis Gas chromatography–Mass spectrometry and liquid chromatography–Ultraviolet absorbance detection–Evaporative light scattering detection. J Chromatogr A 2005, 1062, 227–238.

[15] Cushen, M., Kerry, J., Morris, M., Cruz-Romero, M., & Cummins, E. Nanotechnologies in the food industry–Recent developments, risks and regulation. Trends Food Sci Technol 2012, 24, 30–46.

[16] Dhar, P., Bhardwaj, U., Kumar, A., & Katiyar, V. Poly (3-hydroxybutyrate)/cellulose nanocrystal films for food packaging applications: Barrier and migration studies. Polym Eng Sci 2015, 55, 2388–2395.

[17] Drumright, RE., Gruber, PR., & Henton, DE. Polylactic acid technology. Adv Mater 2000, 12, 1841–1846.

[18] Fortunati, E., Armentano, I., Zhou, Q., Iannoni, A., Saino, E., Visai, L., Berglund, LA., & Kenny, JM. Multifunctional bionanocomposite films of poly (lactic acid), cellulose nanocrystals and silver nanoparticles. Carbohydr Polym 2012, 87, 1596–1605.

[19] Garlotta, D. A literature review of poly (lactic acid). J Polym Environ 2001, 9, 63–84.

[20] Goffin, AL., Raquez, JM., Duquesne, E., Siqueira, G., Habibi, Y., Dufresne, A., & Dubois, P. From interfacial ring-opening polymerization to melt processing of cellulose nanowhisker-filled polylactide-based nanocomposites. Biomacromolecules 2011, 12, 2456–2465.

[21] Gröning, M., & Hakkarainen, M. Multiple headspace solid-phase microextraction of 2-cyclopentyl-cyclopentanone in polyamide 6.6: Possibilities and limitations in the headspace analysis of solid hydrogen-bonding matrices. J Chromatogr A 2004, 1052, 61–68.

[22] Rhim, JW., Hong, SI., & Ha, CS. Tensile, water vapor barrier and antimicrobial properties of PLA/nanoclay composite films. Lwt-Food Sci Technol 2009, 42, 612–617.

[23] Huang, Y., Chen, S., Bing, X., Gao, C., Wang, T., & Yuan, B. Nanosilver migrated into food-simulating solutions from commercially available food fresh containers. Packag Technol Sci 2011, 24, 291–297.
[24] Hauri, JF., & Niece, BK. Leaching of silver from silver-impregnated food storage containers. J Chem Educ 2011, 88, 1407–1409.
[25] Li, H., & Huneault, MA. Effect of nucleation and plasticization on the crystallization of poly (lactic acid). Polymer 2007, 48, 6855–6866.
[26] Lin, N., Huang, J., Chang, PR., Feng, J., & Yu, J. Surface acetylation of cellulose nanocrystal and its reinforcing function in poly (lactic acid). Carbohydr Polym 2011, 83, 1834–1842.
[27] Magnuson, B., Munro, I., Abbot, P., Baldwin, N., Lopez-Garcia, R., Ly, K., McGirr, L., Roberts, A., & Socolovsky, S. Review of the regulation and safety assessment of food substances in various countries and jurisdictions. Food Addit Contam Part A 2013, 30, 1147–1220.
[28] Paneaa, B., Ripoll, G., González, J., Fernández-Cuello, Á., & Albertí, P. Effect of nanocomposite packaging containing different proportions of ZnO and Ag on chicken breast meat quality. J Food Eng 2014, 123, 104–112.
[29] Picard, E., Espuche, E., & Fulchiron, R. Effect of an organo-modified montmorillonite on PLA crystallization and gas barrier properties. Appl Clay Sci 2011, 53, 58–65.
[30] Ray, SS., & Bousmina, M. Biodegradable polymers and their layered silicate nanocomposites: in greening the 21st century materials world. Prog Mater Sci 2005, 50, 962–1079.
[31] Ulery, BD., Nair, LS., & Laurencin, CT. Biomedical applications of biodegradable polymers. J Polym Sci B 2011, 49, 832–864.
[32] Yam KL. ed, The Wiley encyclopedia of packaging technology, John Wiley & Sons, New Jersey, United States, 2010.
[33] Weaire, D., & Hutzler, S. The Physics of Foams, Oxford University Press, NY, USA, 1999.
[34] Eaves, D. Handbook of Polymer Foams. Rapra Technology Limited 2004.
[35] Mills, NJ. Polymer Foams Handbook: Engineering and Biomechanics Applications and Design Guide, Butterworth-Heinemann, 2007.
[36] Lee, ST., Park, CB., & Ramesh, NS. Polymeric Foams: Science and Technology, Taylor and Francis, 2007.
[37] Auras, RA., Lim, LT., Selke, SEM., & Tsuji, H. Poly(lactic acid): Synthesis, Structures, Properties, Processing, and Applications, John Wiley & Sons, 2010.
[38] Song, J., Kay, M., & Coles, R. Bioplastics, Coles R, Kirwan M.J., ed, Food and Beverage Packaging Technology, 2nd ed, Oxford, UK, Wiley-Blackwell, 2011, 295–319.
[39] Klempner, D., & Sendijarevic, V. Polymeric foams and Foam Technology, Hanser, Munich, Germany, 2004.
[40] Lee, ST., & Ramesh, NS. Polymeric foams: Mechanisms and Materials, CRC Press, Taylor & Francis Group, 2004.
[41] Arif, S., Burgess, G., Narayan, R., & Harte, B. Evaluation of a biodegradable foam for protective packaging applications. Packag Technol Sci 2007, 20, 413–419.
[42] Willett, J.L., & Shogren, R.L. Processing and properties of extruded starch/polymer foams. Polymer 2002, 43, 5935–5947.
[43] Preechawong, D., Peesan, M., Supaphol, P., & Rujiravanit, R. Preparation and characterization of starch/poly(l-lactic acid) hybrid foams. Carbohydr Polym 2005, 59, 329–337.
[44] Lee, SY., Chen, H., & Hanna, MA. Preparation and characterization of tapioca starch–Poly(lactic acid) nanocomposite foams by melt intercalation based on clay type. Ind Crops Prod 2008, 28, 95–106.
[45] Guan, J., Eskridge, KM., & Hanna, MA. Acetylated starch-polylactic acid loose-fill packaging materials. Ind Crops Prod 2005, 22, 109–123.

[46] Ajioka, M., Enomoto, K., Yamaguchi, A., Suzuki, K., Watanabe, T., & Kitahara, Y. US Patent 5,447,962. 1995.
[47] Zhou, C., Ma, L., Li, W., & Yao, D. Fabrication of tissue engineering scaffolds through solid-state foaming of immiscible polymer blends. Biofabrication 2011, 3, 4.
[48] Tang, L., Zhai, W., & Zheng, W. Autoclave preparation of expanded polypropylene/poly(lactic acid) blend bead foams with a batch foaming process. J Cell Plast 2011, 47, 5, 429–446.

Siddharth Mohan Bhasney, Prodyut Dhar and Vimal Katiyar

8 Polymer blends for sustainable food packaging

8.1 Introduction

Due to the increasing cost and limited petroleum resources, polymeric products developed from the petroleum-based resources cause severe environmental hazards, which require immediate replacement with biobased and biodegradable polymers. These have resulted in developing a keen interest for the development of biodegradable packaging materials due to their eco-friendly nature, improved safety, nontoxicity in addition to declined unit cost and economic viability. In comparison to synthetic nondegradable polymers, biobased polymers have poor gas permeability, low thermal stability and reduced processability through industrial viable extrusion process. To overcome such challenges, blending of polymers has shown to enhance its physicochemical and structural properties, to improve its processing capabilities as well as is proved to be a cost-effectiveness process. Among the all attributes, blending of polymers or with the introduction of additives (such as adding color or compatibilizers) has shown to improve the coefficient of friction, promoting adhesion, enhancing stability and attaining easier operating features. For simple blending, all the raw ingredients are mixed well in an extruder in which the resin is converted into a film, thread or coating. The complex blending is usually carried out through industrial-scale extruders, which have specialized screw designs or customized integrated equipment that are required for attaining the desired properties. These machines are incorporated into numerous mixing modes, namely, flow rearrangement or higher stress levels for breaking the particles, that is, dispersive mixing, which leads to improved dispersion of both the polymers. Alternative routes such as complex shear and elongation flow might also be applied for achieving the optimum mixing between the two polymers. The property of the finally achieved blended material depends upon the flow, stress history and also onto the thermodynamics of mixing the two components. The minor components form a different dispersed phase and the major one forms a continuous phase in most of the polymer blends that are immiscible [1].

8.2 Biopolymer

Over the past few years, biopolymer is among the fastest growing material sectors that have been widely used for the fabrication of polymers from the biomass-derived

https://doi.org/10.1515/9783110648034-008

renewable sources and have potential to replace fossil fuel-based polymers (Table 8.1, Figure 8.1). The prime factors for that enhanced utilization of biopolymers are as follows: (1) limitations of fossil fuel resources; (2) enhanced fuel prices; (3) direct adverse effect of fossil fuels onto the climate change and (4) waste discarding problem generated by the fossil fuels derived polymers [2].

Table 8.1: The major processing routes to potential biobased products.

Processing techniques	Application
Coextrusion	Packaging film
Cast film	Packaging film
Thermoformed sheets	Trays and cups
Blown film	Packaging plastics bags
Injection blow molding	Salad pots, cutlery, drinking beakers, cups, plates, drinking bottle and trays

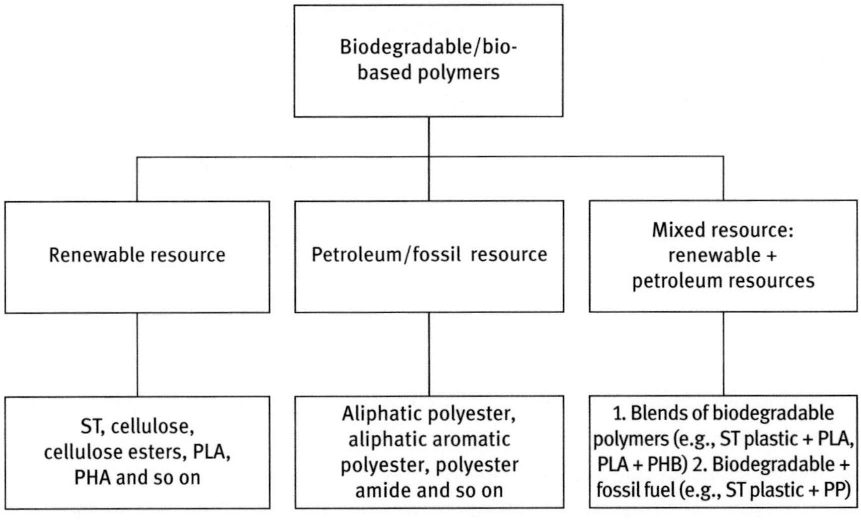

Figure 8.1: Broad classification of biodegradable and biobased polymer.

Biopolymers are named under the organic polymers. These polymers are naturally derived and have been found in nature for around billions of centuries developed through biological processes and comprise of, namely, starch, proteins, peptides, deoxyribonucleic acid and ribonucleic acid. They are generally made up of long chain molecular subunits, which are eco-friendly, biodegradable and nonhazardous biomaterials. These biopolymers can be obtained from several renewable (natural) resources, which are products and byproducts of many microbes, plants or organisms, which could either be synthesized from the biotic materials such as pentose and hexose sugars, starch, natural fats or oils and so on.

8.2.1 Source and description of biobased polymers

Depending upon the source of origin and process for productions or extraction, biobased polymers may be divided into three main categories:

Primary feedstock: This consists of polysaccharide-based, protein-based bioplastic such as rice, corn, sorghum, sugarcane, beet and so on.

Secondary feedstock: This involves the use of renewable biobased raw materials. This category involves the development of poly(lactic acid), polymer of lactic acid, which can be produced through fermentation of carbohydrate.

Tertiary feedstock: This class includes the production of polymers through the use of microorganism or genetically modified bacteria such as poly(hydroxyl alkanoates).

The potential applications of the naturally derived biodegradable polymers as food packages products are relatively limited due to their poor structural and barrier properties. Because of that, these natural biomaterials are frequently blended with other synthetic polymers or are chemically modified through various novel strategies, which lead to its wider applications and have been discussed in the subsequent section [2].

8.3 Biodegradable polymer-based blends

Blending of the two or more biodegradable materials, namely, starch, polyvinyl alcohol, polyhydroxylbutyrate (PHB) and so on, leads to the reduction of brittleness of biodegradable polymers. In this regard, the dispersed fillers act as nucleating as well as reinforcing agents, which lead to improvised thermomechanical properties. Other relevant properties of the prepared nanocomposites such as morphology, processing conditions, glass transition temperature and crystallinity are significantly influenced by the characteristic property of the blends [3, 4]. Introduction of different fillers from such wide pool of renewable materials can also be said as a novel strategy to tailor the properties of PLA while upholding its biodegradable nature [5].

For potential application of the biodegradable polymeric blends in food-packaging products, it should be able to protect the food substances from the light, water or any other related chemically oxidative process [6]. As there is a great significance of the crystalline phase on to the mechanical and permeation properties, therefore, in the prior part, more focus has been given on enhancing the crystallinity of the polymer material. For example, in case the of enhancing PLA crystallinity, the incorporation of highly crystalline PHB into the PLA matrix is attempted by melt blending, which improves the mechanical properties of compounded blend [7]. In addition, the similar melting temperature of PHB and PLA has made the blending much easier, resulting in enhancing the barrier properties especially for oxygen and water vapor. But this leads to a decrease in the inherently high transparency of the PLA, which is also a much considered issue for the fabrication of the polymeric composites intended for food packaging applications. Also, for preventing the food stuffs from the ultraviolet light, transparent films of PHB are considered for achieving an upper edge over PLA; however, PHB has been proposed as a good potential in the short-term food packaging [8, 9].

As one of the attempt to use poly(lactic acid) and poly(hydroxybutyrate) blend for food packaging, it has been plasticized with a natural terpene D-limonene (LIM) using melt compounding to improve the PLA crystallization and also to reduce the glass transition temperature to develop flexible film. Moreover, functional properties were studied by oxygen permeation and water-resistant measurements along with structural and surface properties. The characteristic bands corresponding to PLA and PHB and their molecular interaction were shown by Fourier transform infrared radiation (FTIR) spectra. Py-GC/MS showed the characteristics peak of D-limonene as well as the thermal degradation products of PLA and PHB. It has been found that the D-limonene amount was higher in PHB phase than that in PLA. PHB produced a reinforcing effect in the PLA matrix, which subsequently induced improved oxygen barrier properties and the resistance to the permeation of surface water across it. In another study, introduction of D-limonene had a plasticization impact onto its mechanical properties, which led to enhancement in the elongation at break. Further, the composting conditions confirmed that PHB delays the PLA composting, whereas D-limonene tried to enhance the same. Hence, for the ternary combination of PLA–PHB–terapene-D-limonene (LIM) film, properties such as structural, thermal, barrier and mechanical behavior were found out to be the best [10].

8.3.1 PLA–PCL blends

PCL is another class of petroleum-derived polymers but undergoes degradation within a stipulated time period of ~180 days; therefore they are classified as biodegradable polymers. Due to its hydrophobic nature, both PLA and PCL have improved compatibility during blending, which leads to enhanced structural and barrier properties, the extent of which depends upon the loading fraction of

PCL in PLA matrix. PCL inherently has high elongation behavior; therefore when introduced into PLA it overcomes its brittle nature along with improvement in thermal behavior [11]. It has been reported that incorporation of PCL content as high as ~30 wt% has shown to improve the elongation behavior of PLA by ~85% as well as its crystallinity [12]. Crosslinking of PLA with other biodegradable polymers such as PCL and PHB through thiolene-based functionalization has shown to improve their compatibility along with improvement in the glass transitions temperature [13]. Laminated multilayered composites of PLA and PCL have been shown to have significantly improved structural properties with the presence of multiphasic porosity, which finds potential applications in development of scaffolds as well as drug release agents [14]. Therefore, blending of PCL into PLA has been shown to improve the physicochemical, structural properties of PLA, especially its poor elongation behavior along with its inherent biodegradable nature, which finds potential applications in various sectors such as packaging, tissue engineering and drug delivery.

8.3.2 PLA–PBS blends

Studies on blending of PLA with poly(butylene succinate) (PBS) at different compositions were prepared by using a twin screw compounder. Further, the investigations of the physical, thermal, rheological and structural properties and morphological analysis of developed films were carried out to understand the dispersion of blends in context with packaging applications. The presence of clear phase difference trend could be observed depending on the blend composition especially at higher concentrations. These led to declination of the tensile strength, modulus and percentage (%) elongation at break in PLA. The rheological behavior showed attenuated results of the blend possessing PBS concentration ≤20% by weight. Lastly, the reduction in the brittleness of PBS-blended PLA, is favourable for its use in the replacement of plastics for the packing applications [15].

8.3.3 PLA–PBAT blends

The preparation of poly(butylene adipate-*co*-terephthalate) (PBAT) blend with PLA is generally performed by utilizing the solvent-casting method. After the characterization of the blended polymer, it forms a homogenous blend film. Moreover, the FTIR and differential scanning calorimetry analysis come with the knowledge that blends of PLA and PBAT result in the formation of homogeneous films. The blending of PLA with a small amount of PBAT results in a significant increase in flexibility of PLA films without varying its mechanical properties. One study shows that the blending 3.92 gm of PLA with 0.08 gm of PBAT had notable UV screening effect without affecting its film transparency. However, a proportion of blends greatly influences the water vapor and oxygen transfer rate of films, which is a crucial factor for acting as a good packaging material. Wang et al. reported the effects of blends films on storage of potatoes and green onions, which show an antifogging effect, resulting in a reduced quality changes that help in preventing water condensation [16].

8.3.4 PCL/chitin–chitosan-based blends

Melt-blending techniques were used to develop the blends of PCL with chitin, a biobased composites. Chitin contents in biobased composites generally do not affect the melting or crystallization temperature. In contrast, blend of chitin with PCL has a notable effect on crystallinity of PCL. In addition to melt-blending techniques, solvent-casting method can be used to develop biobased composites for food packaging [17].

8.3.5 Cellulose-based blends

Literature showed that the combined usage of polymer materials exhibited as a substitute for packaging several food and food products with better quality retaining properties. Among all available blends, paper board, a cellulose-based packaging material, is one of the most widely used renewable packaging material. In addition, various cellulose derivatives such as cellulose acetate are widely used in packaging of fresh commodities, vegetables, bakery products and so on, with the utilization of chitosan–cellulose- and PCL-based laminate films, which have permeability as that of LDPE. LDPE as packaging material has a wide application in modified atmospheric packaging (MAP). Due to improved barrier properties of chitosan–cellulose- and PCL-based laminate films, the use of such polymeric composite films directs their utilization in MAP for increasing the shelf life of fresh products (banana, mango, apple, pear, sapotas, guava, etc.). Coating of the PLA onto paperboards has shown promising properties as a replacement to PE-coated paperboards for the development of paper containers with high moisture barrier. Furthermore, the use of cellophane with poly(vinylidene chloride) or nitrocellulose provides better barrier properties, which can be used extensively in the packaging of bakery products, milk products, processed meat and so on [11].

8.4 Blends of other fossil fuel-based plastics and biobased plastics

Biobased polymers, like PLA, PHB, PCL, PHA, starch and so on, are attempted to blend with a variety of fossil fuel-based nonbiodegradable polymers such as PET, PS, PP, PVC, PC, PE and their copolymers to achieve a partially biodegradable product. Presently available partially biodegradable materials are starch–PE blends, are used in the packaging and cutlery industries. There are some drawbacks of PLA in terms of its properties, such as lower heat deflection temperature, highly brittle nature, high production cost, poor processability along with low barrier properties due to which the biopolymers are blended with nonbiodegradable polymers to improve their inherent characteristics as listed in Table 8.2.

Table 8.2: The major processing routes to representative plastic articles.

Processing route	Product examples
Film-blowing process	Packaging film
Other extrusion based processing:	Packaging films (wrappers and so on)
Cast-sheeting process	Processing of tubes, hoses, pipes
Pipe extrusion	Films for thermoforming, vacuum forming and so on
Sheet extrusion	Flexible film
Tubular blown film extrusion	Multilayered films
Coextrusion	
Thermoforming	Trays and cups, automobile parts
Injection blow molding	Bottles
Compression and transfer molding	Machinery parts
Rotational molding and sintering	Large hollow plastic parts and products
Plastic foam molding	Plastic-based foams

Medical and packaging applications require improved thermal, mechanical, chemical and barrier properties, which may be obtained by blending biopolymers with synthetic petroleum-based polymers. However, the preparation of such blends should incorporate a higher percentage of biopolymers than the use of conventional polymers to enhance the biodegradable nature of the composites. Recently, in some studies the biobased polymers are blended with polyolefins, vinyl polymers, vinylidene polymers, copolymers, elastomers, rubbers and so on. It is known to us if polyethylene (PE) or polypropylene (PP) once blend with PLA, it can provide higher toughness, good impact strength and are of low cost and are useful for commercial packaging applications. However, the dissimilarity in the chemical structures of polyolefins and PLA leads to the immiscible behavior of blends, which causes weak interfacial adhesion, poor mechanical properties and distinct particle interfaces with different morphologies [18].

8.4.1 PLA–LDPE blends

Raghavan and Emekalam [19] studied the effect of introducing starch as an additive to the PLA/PE-based blends especially to understand its effect on degradation behavior. The starch additives were shown to enhance the acid degradability of the blend. Mechanical properties of PLA/LDPE blend were investigated by using cotton by-products as fillers so as to reduce the high production cost of PLA, and the results indicated that the filler materials increased Young's modulus but there was a reduction in strength and elongation. It is noteworthy to mention that food package using PLA/LLDPE blend films have been developed at CoE-Suspol, IIT Guwahati, India, as shown in Figure 8.2.

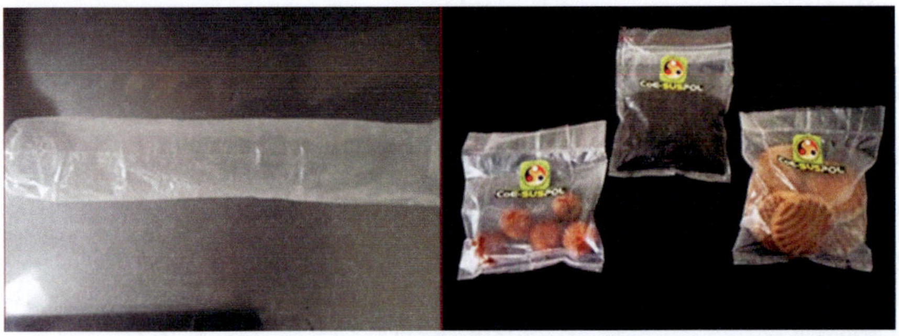

Figure 8.2: Food packaging in PLA/LLDPE films developed at CoE-Suspol, IIT Guwahati, Guwahati.

PLA/LLDPE (linear low-density polyethylene) polymer blends were prepared by Balakrishnan et al. [20] with low LLDPE content (<20 wt%) and their mechanical and thermal properties were obtained. A reduction in tensile properties and poor blend compatibility were observed. Hamad et al. [21] also reported the poor compatibility of PLA/LDPE blends, wherein the blends were processed through a single screw extruder. The compatibility of PLA and LDPE blends was determined by the morphological studies for understanding the dispersion of LDPE fractions in PLA, which resulted in variation of the mechanical and rheological properties of the fabricated blends. Compatibilizers such as PLA-*co*-PE, PE-*graft*-glycidyl methacrylate (PE-*g*-GMA), styrene–ethylene–butylene–styrene-grafted maleic anhydride (SEBS-*g*-MA) and PE-*g*-MA were used to improve the miscibility and thereby the tensile properties of PLA/PE blends. The presence of compatibilizers helps in the reduction of the particle size of dispersed polymeric phase and provides strong interfacial interaction between the two phases, resulting in enhanced mechanical properties of the blend.

In another study, the structural properties of PLA/PE blends were studied by Anderson and Hillmyer [22], which were compatibilized by PLA-*co*-PE. It was found that blending PLA with PE improved the impact strength when compared to pure

PLA, which was further enhanced by adding compatibilizers, whereas the other mechanical properties such as tensile strength, modulus and so on were not found to improve significantly in comparison to PLA. Kim et al. [23] reported the use of PE-chemically grafted-GMA process for improving tensile properties of PLA/LDPE blends resulting from the chemical reaction between PLA and PE-*g*-GMA, which causes a strong interfacial interaction between PLA and LDPE. Singh et al. [24] used PE-*g*-GMA for the preparation of a blend containing 20 wt% PLA and 80 wt% LLDPE in which the optimal tensile strength and elongation at break were obtained by using ~4 wt% of the compatibilizer, which could be attributed to the improved interaction between the compatibilizers in the blend [25].

8.4.2 LDPE–rice/potato starch blends

The preparation of LDPE and rice/potato starch-based blends through extrusion process in the presence of varying concentrations of water was evaluated before and after storage for a certain period of time to understand the effect of change in the structural properties, gas or water permeability and biodegradability. The mechanical properties of LDPE/starch blends were undesirable, which were due to the higher percentage of starch (>30% w/w), whereas water and gas permeation rate increased with increasing starch content in the blend. However, the biodegradability rate was found to improve on adding more than 10% (w/w) starch [26].

8.4.3 Recycled PET/PLA–PET/chitosan-based blends

The miscibility and degradability of films of recycled PET (polyethylene terephthalate)/PLA and PET/chitosan blends prepared through extrusion at 250°C were examined and studied in detail. XRD and infrared spectroscopy-based studies depicted weak interfacial adhesion between PET matrix and biodegradable polymers. The semicrystalline character was affected by the mobility of biopolymer fraction and its quenching protocol, which helped in the recrystallization of R-PET during the extrusion reprocessing based on the processing conditions. The polymer–biopolymer blend miscibility was independent of the PET matrix, which attained saturation with 10 wt% PLA, whereas an adequate miscibility was obtained for all compositions of chitosan. SEM analysis indicated uniform distribution of green polymers into the polymer matrices, but the amount and type of biopolymer resulted in varying dimensions of the agglomerates and their dissolution. When compared to PET-based PLA blends, the interaction between PET and chitosan favored enhanced degradation rate as indicated by accelerating, weathering tests. The PET/chitosan blend with 95:5 weight ratio was observed to have the best performance where an estimate for its degradation time was found to be about 45 years [27].

8.4.4 PLA/PP blends

PLA and PP are partially compatible, which can be processed only through extrusion-based processes when the two thermoplastics are in melt phase. The morphology of the blend having 50 wt% PLA showed ill-defined interfaces, which were similar to those obtained in the compatibilized PLA/PE systems. Besides, PLA/PP blend fibers are more resistant to hydrolysis than pure PLA. PLA/PP blends were processed through a single screw extruder in which the interfacial tensions showed higher compatibility of blend when compared to other PLA/synthetic polymer systems. Rheological studies revealed that fabricated blend had low viscosity when compared to pure PP, leading to the improved extrusion ability. A variety of compatibilizers such as PP-*g*-MA and PLA/PP/PP-*g*-MA were used to improve the interfacial compatibility between 90 wt% PLA and 10 wt% PP. The effect of the compatibilizers were investigated by thermal, mechanical and morphological properties. Thermal stability of PLA/PP/PP-*g*-MA blend was better than pure PLA, uncompatibilized blends and another compatibilized blend (PLA/PP/PP-*g*-GMA). This behavior was attributed to finer and more homogenous distribution of PP phase in PLA matrix [25].

8.4.5 PLA/polycarbonate (PC) blends

In this study, PC, PBAT and PLA blends provided significantly improved percentage elongation, tensile strength and impact strength due to the presence of dicumyl peroxide (DCP) as a radical initiator. DCP induced the reaction between PBAT and PLA, resulting in a good dispersion. On the other hand, PC tended to locate near the new phase (PBAT–PLA) due to its good compatibility, leading to a multidispersed phase of PLA/PBAT and PLA/PC. The newly developed morphology contributed to enhanced the ductility of the fabricated material. The tensile properties of PLA/PBAT and PLA/PC were enhanced due to the considerable role of DCP [25].

8.4.6 PLA/polystyrene (PS) blends

PLA/PS blends are significantly reported where interfacial interactions between phases lead to the good compatibility between them and improve the thermal properties. This is due to the presence of physical bonding between the lone paired of electrons (n) for the electronegative carbonyl groups in PLA and π electrons of phenyl rings in PS (n–π bond). It happened by shifting of carbonyl absorption band (near 1,760 cm^{-1}) of pure PLA toward the less wave numbers in the blend. This interaction resulted in a kind of poor compatibility, which led to a higher thermal stability of the blend compared with pure PLA. Compatibility of PLA/PS polymer blends was estimated by studying tensile properties and was compared with that of other PS/biodegradable systems including PCL/PS and PBS/PS prepared using the extrusion process. A good compatibility was recognized in PCL/PS and PBS/PS blends and was compared with PLA/PS. Effect of PLA percentage resulted in an increment of rheological properties, whereas the viscosity of the PLA/PS blend was decreased [25].

8.4.7 LDPE–starch blends

In this study, two different methods, that is, solvent casting followed by compression at high temperature and extrusion were used for the fabrication of LDPE and starch blend films in which loading of starch was varied from 2.5–50 wt%, which led to a significant variation in its properties. The films showed promising results for potential applications as food packages and therefore determination of the overall migration of the films in the presence of different food stimulants was found to be within the limits as per the standard legislations. Furthermore, the blended films of LDPE/starch show enhanced biodegradability rate due to the presence of starch, which increased its degradation rate [28].

8.5 Conclusion

Biodegradable polymers and its blends are sustainable and environmentally benign plastics, and after their potential applications undergo complete fragmentations and bioassimilation under composting conditions within a stipulated time period. These prevent the accumulation of such polymers in the environment unlike the petroleum-derived polymers, which are known for their potential environmental toxicity and global warming issues. Development of polymeric blends utilizing the biopolymers or biobased polymers leads to improvement in the physicochemical, structural and barrier properties, keeping the inherent characteristics of biodegradability intact. Processing of such biodegradable polymeric blends through the industrially scalable extrusion-based approaches will lead to the development of polymeric products for potential applications in various sectors, which will play an important role to replace the petroleum-derived polymers.

References

[1] Ebnesajjad, S., & William, Andrew. Plastic Films in Food Packaging: Materials, Technology and Applications, The Boulevard, Langford Lane, Kidlington, Oxford, OX5 1B 225 Wyman Street, Waltham, MA 02451, USA, 2012.

[2] Niaounakis, M., & Andrew, William. Introduction to Biopolymers-Rationale for Use of Biopolymers, Biopolymers: reuse, recycling, and disposal, The Boulevard, Langford Lane, Kidlington, Oxford, OX5 1B 225 Wyman Street, Waltham, MA 02451, USA, 2013.
[3] El-Hadi, A., Schnabel, R., Straube, E., Muller, G., & Henning, S. Correlations between degree of crystallinity, morphology, glass temperature, mechanical properties and biodegradation of poly (3-hydroxyalkanoate) PHAs and their blends. Polym Test 2002, 21, 665–674.
[4] El-Hadi, A., Schnabel, R., Straube, E., Müller, G., & Riemschneider, M. Effect of melt processing on crystallization behavior and rheology of poly (3-hydroxybutyrate) (PHB) and its blends. Macromo Mater Eng 2002, 287, 363–372.
[5] Ohashi, E., Drumond, WS., Zane, NP., Barros, PWDF., Lachtermacher, MG., Wiebeck, H., & Wang, SH. Biodegradable Poly (3-hydroxybutyrate) Nanocomposite. Macromo Symp 2009, 279, 138–144.
[6] Castro López, MDM., De Dicastillo, CL., Vilariño, JML., & Rodríguez, MVG. Improving the capacity of polypropylene to be used in antioxidant active films: Incorporation of plasticizer and natural antioxidants. J Agric Food Chem 2013, 61, 8462–8470.
[7] Wang, J., & Mao, Q. A novel process control methodology based on the PVT behavior of polymer for injection molding. Adv Polym Tech 2012, 32, 474–485.
[8] Bucci, DZ., Tavares, LBB., & Sell, I. Biodegradation and physical evaluation of PHB packaging. Polym Test 2007, 26, 908–915.
[9] Khosravi-Darani, K. Application of poly (hydroxyalkanoate) in food packaging: Improvements by nanotechnology. Chem Biochem Eng Qtr 2015, 29, 275–285.
[10] Arrieta, MP., & López, J. Hernández A and Rayón E, Ternary PLA–PHB–Limonene blends intended for biodegradable food packaging applications. Eur Polym J 2014, 50, 255–270.
[11] Peelman, N., Ragaert, P., De Meulenaer, B., Adons, D., Peeters, R., & Cardon, L. Van Impe F and Devlieghere F, Application of bioplastics for food packaging. Trends Food Sci Techno 2013, 32, 128–141.
[12] Ferri, JM., Fenollar, O., Jorda-Vilaplana, A., García-Sanoguera, D., & Balart, R. Effect of miscibility on mechanical and thermal properties of poly (lactic acid)/polycaprolactone blends. Polym Int 2016, 65, 453–463.
[13] Belkhir, K., Jegat, C., & Taha, M. Thiolene grafting from polylactic acid, polycaprolactone, and polyhydroxybutyrate. React Funct Polym 2016, 101, 82–89.
[14] Semba, T., Kitagawa, K., Ishiaku, US., & Hamada, H. The effect of crosslinking on the mechanical properties of polylactic acid/polycaprolactone blends. J Appl Polym Sci 2006, 101, 1816–1825.
[15] Bhatia, A., Gupta, RK., Bhattacharya, SN., & Choi, HJ. Compatibility of biodegradable poly (lactic acid) (PLA) and poly (butylene succinate) (PBS) blends for packaging application. Korea Aust Rheo J 2007, 19, 125–131.
[16] Wang, LF., Rhim, JW., & Hong, SI. Preparation of poly (lactide)/poly (butylene adipate-co-terephthalate) blend films using a solvent casting method and their food packaging application. LWT – Food Sci Techno 2016, 68, 454–461.
[17] Pawar, PA., & Purwar, AH. Biodegradable polymers in food packaging. Am J Eng Res 2013, *2*(5), 151–164.
[18] Auras, RA., Lim, LT., Selke, SE., & Tsuji, H. Poly (lactic acid): synthesis, structures, properties, processing, and applications, Vol. 10, John Wiley & Sons, Inc, Hoboken, New Jersey, 2011.
[19] Raghavan, D., & Emekalam, A. Characterization of starch/polyethylene and starch/ polyethylene/poly (lactic acid) composites. Polym Degrad Stab 2001, 72, 509–517.
[20] Balakrishnan, H., Hassan, A., & Wahit, M. Mechanical, thermal, and morphological properties of polylactic acid/linear low density polyethylene blends. J Elastomers Plast 2010, 42, 223–239.

[21] Hamad, K., Kaseem, M., Ko, YG., & Deri, F. Poly (lactic acid)/low density polyethylene polymer blends: Preparation and characterization. Asia-Pacific J Chem Eng 2012, 7, S310–S316.
[22] Anderson, K., & Hillmyer, M. The influence of block copolymer microstructure on the toughness of compatibilized polylactide/polyethylene blends. Polym 2004, 45, 8809–8823.
[23] Kim, YF., Choi, CN., Kim, YD., Lee, KY., & Lee, MS. Compatibilization of immiscible poly (l-lactide) and low density polyethylene blends. Fiber Polym 2004, 5, 270–274.
[24] Singh, G., Bhunia, H., Rajor, A., & Choudhary, V. Thermal properties and degradation characteristics of polylactide, linear low density polyethylene, and their blends. Polym Bull 2011, 66, 939–953.
[25] Hamad, K., Kaseem, M., & Ko YG and Deri, F. Biodegradable polymer blends and composites: An overview. Polym Sci Serv 2014, 56, 812–829.
[26] Arvanitoyannis, I., Biliaderis, CG., Ogawa, H., & Kawasaki, N. Biodegradable films made from low-density polyethylene (LDPE), rice starch and potato starch for food packaging applications: Part 1. Carbohydr Polym 1998, 36, 89–104.
[27] Torres-Huerta, AM., Palma-Ramírez, D., Domínguez-Crespo, MA., Del Angel-López, D., & De La Fuente, D. Comparative assessment of miscibility and degradability on PET/PLA and PET/chitosan blends. Eur Polym J 2014, 61, 285–299.
[28] Raj, B., & Udaya Sankar, K. Low density polyethylene/starch blend films for food packaging applications. Adv. Polym. Techno.: J Polym Process Inst 2004, 23, 32–45.

Narendren Soundararajan and Vimal Katiyar

9 Biobased biodegradable polymers in food packaging: regulations and legislations

9.1 Introduction

In recent days, there has been a growing concern regarding the safety of food consumers due to the use of engineered nanoparticles (ENPs) as additives in packaging materials to improve the properties. Most packaging materials including biobased polymers contain chemicals that have the potential to migrate upon interaction with food products. These chemicals can be biobased monomers used in plastic productions, residual levels of chemicals used in the refining/processing/manufacture of the materials, any additives like plasticizers, antioxidants, ENPs and so on, added to the formulation to provide the required characteristics to these materials.

Migration is a diffusion process that releases materials from one type of medium to another via a concentration gradient that follows Fick's diffusion law [1]. The substance that is migrating is known as *migrant*. There are several factors that affects migration characteristics including time of contact, temperature of storage, concentration, properties of the material and the type of interactions. In general, migration as a whole cannot be referred to as a health issue but is considered as a legal issue according to the legislations of developed countries.

The ENP shows several uncertainties in migration; thus, a strict rule has been implemented for the usage of nanoparticles in food contact materials (FCMs). In every country, governing authorities have formulated their own approaches for the use of ENPs in FCM, which are some guidelines or a regulation that is suitable for the region to adopt. The European Commission (EC) has formulated collective legislation to safeguard the consumers from the effects of migration that comes from food-packaging materials. European Communities (CEC) in 1976 proposed the first directives for the standard analytical methods and detection limits for various toxic migrants that are possible to migrate during food contact, which is supplied among the state members of EU.

The legislative directives have formulated two methods, overall migration (OM) and specific migration (SM), to test and prove the occurrence of migration from FCMs. OM is "referred as the sum of all (frequently not known) mobile substances of packaging released per unit area of package under the influence of specific predetermined conditions." SM "is only related to a migration of specific, known substance" [2]. Test for OM is easy to perform than that for SM test. If the OM value is within the limit as mentioned by the directives, then there is no need of SM test as the substance that has migrated is too low to cause a public health hazard.

https://doi.org/10.1515/9783110648034-009

The concept of overall migration limit (OML) has been introduced to regulate the total quantity of a substance that has migrated to the food packaging and to minimize the SM testing that is needed to prove that the packaging material is appropriate for use in food packaging. In Directive 90/128/EEC, the OML is set to 10 mg/dm^2 (60 mg/kg foods) for migration of chemicals, unpolymerized monomers and other additives that are used in plastic manufacturing industries approved for food packaging. Substances that show OM limit value lower than the intended value (10 mg/dm^2) do not essentially prove that the material is harmless to use for food contact application, as the OM cannot recognize the chemical and toxic nature of the migrant species. On every revision, the regulatory body publishes an updated positive list of materials and chemicals for use in FCMs. Only those materials that appear in the positive list can be commercially used for the application in the intended field.

The directive in each country follows different procedure and have different OML limits of their own. Even though all the directives mention the use of food simulants as the testing material for migration, but the use of food materials consume more time and the testing procedure is complex. The food simulants that simulate the natural properties of food are widely used as they cover a wide food range for complete assessment of human exposure compared to consumption data. In some cases, the usage of fatty acids as food simulants becomes worst-case scenario as it is difficult to test the materials analytically. In some studies, the available food products such as chicken and turkey meat have been used as testing matrices for testing the antimicrobial effect of active packaging [3, 4]. The real food matrix can benefit packages that are produced for specific food application. In other cases, the usage of proper food simulant is applied for reliable results.

As previously mentioned, migration follows Fick's law of diffusion and is one of the main mechanism of transfers of chemical species from FCMs to food. The other possible mechanism can be absorption. The chemical species that come in contact with the food material without affecting the physiochemical nature of the food follows Fick's first and second law as mentioned in eqs. (9.1) and (9.2) [5].

$$\text{Fick's first law: } F = -D_p(\delta C_p/\delta x) \tag{9.1}$$

$$\text{Fick's second law: } \delta C_p/dt = D_p(\delta^2 C_p/\delta x^2) \tag{9.2}$$

where F is the rate of transport per unit area of the polymer; D_p is the coefficient of diffusion of migrant in the polymer (cm^2/s); C_p is the concentration of migrants in the polymer matrix (mg/g); x is the interface between polymer and food (cm) and t is elapsed time (s). When migration is diffusion controlled, the following eq. (9.3) can be used [6]:

$$M_t = 2C_o\rho\ (D_p t/\pi) \tag{9.3}$$

where M_t is the total migrants from the polymer in time, t (s); C_o is the concentration of initial migrant in the polymer (mg/g); ρ is the polymer density (g/cm^3); D_p is the coefficient of diffusion of the migrants in the polymer (cm^2/s) and t is the total package lifetime in seconds.

9.2 Nanoparticles in food packaging

The usage of ENPs in packaging applications pose tremendous serious harm to humans and the surrounding environment because of its high toxicity, flexibility and bioaccumulation. The consumers should be informed about all risk associated with the use of a material that is produced by nanotechnology and should have the freedom to accept nanotechnology-related products with proper information on the product. In addition, the test for the risk level to human needed to be carried out in worst-case condition, if the particular material poses an unacceptable risk, a risk management strategy has to be established to alleviate the risk. Migration studies and in vivo toxicity studies are the key tool to identify the risk level to humans on assimilation of ENPs. Based on the results of these studies, the concerned regulatory authority will permit or prohibit the use of the product. The acceptance of tin nanoparticles (TiN) by the European Food Safety Authority (EFSA) for its usage in PET bottles is a success to such risk assessment process. EFSA has allowed concentrations up to 20 mg/kg to use in these bottles as it shows improved oxygen barrier properties when incorporated into PET. Another material in the list is engineered silver nanoparticles that migrate as Ag ions under accepted level permitted by EC (EFSA, 2008; European Commission, 2011), which express its antimicrobial function on contact with food and increase food shelf-life.

9.3 Experimental migration studies

9.3.1 Migration studies on PLA

Migration studies in PLA show that the diffusion of lactic acid from packaging material is lower than the level that is present in the common food ingredients. When PLA film is tested for 10 days at 43 °C in 8% ethanol solution and olive oil, the OM was 0.85 and 0.15 mg/dm^2 in the respective simulants. The standard test procedure was followed as directed by FDA during the migration studies. The migrant consisted of lactic acid, lactoyl lactic acid (acyclic dimer), trimer, and lactide (cyclic dimer). Dimers and hydrolyzed oligomers that migrated into the aqueous systems were the common food ingredients that are listed safe for use [7].

In a study, PLA films were tested for a duration of 15 days at 40 °C under food simulants, such as 10% ethanol and 95% ethanol. In both the simulants, it is exactly proved that the value of migration was lower than the quantity of lactic acid that is present in the average human intake daily (22 mg per day). In another study [8], the OM of a PLA sheet was less than 1 mg/dm^2 in three food simulants such as 3% acetic acid, 15% ethanol and olive oil, for 10 days at 40 °C, and in *iso*octane for 30 min at 40 °C [9]. Different types of PLA sheets were subjected to migration studies under standard test conditions. The oligomers of lactic acid and lactide content migrated into the simulants were determined using liquid chromatography/mass spectroscopy. The total migration level of lactic acid from PLA films is 0.028–1.5 mg/dm^2 at 40 °C for 180 days, where the OM value increased to 0.073–284 mg/dm^2 at 60 °C for 10 days, which was due to decomposition of PLA under the test condition.

9.3.2 Migration studies on bionanocomposites

Many studies were carried out to know the effect of ENPs migration from food packaging both in simulants and real food as matrices. For example, Ag and Ag-based nanoparticles migration were studied on polymers like PVC, PE, low-density polyethylene (LDPE), PP and biodegradable polymers such as PLA nanocomposite [10] and starch-based biopolymers [11]. The main concern for such studies was the addition of ENPs in the polymer matrices to improve its inherent properties. In some cases, ENPs like montmorillonite (MMT) are used for increasing the biodegradability of starch-based polymers and improving oxygen barrier properties by the addition of CNC to PLA [10].

Recycled PET flakes were tested for migration [12]. The migration solutions obtained by the contact of samples for 3 h with the simulants such as water, acetic acid 3%, ethanol 10% and ethanol 95% at 70 °C were screened by ultra performance liquid chromatography – ultraviolet/mass spectroscopy technique. It was observed that identification was difficult due to the lack of high-performance liquid chromatography (HPLC-MS) libraries, and not all migrants could be identified. PET monomers, diethylene glycol (DEG), isophthalic and terephthalic acid, and PET dimer and trimer and UV absorbers could be found beside additives that originated from the contact with different plastic materials such as polyolefins. The first three mentioned substances were quantified and showed very low concentrations.

After screening several recycled and virgin PET materials, the samples were heated up to 200 °C and then analyzed by head space-gas chromatography/mass spectroscopy (HS-GC/MS). In addition, PET flakes were extracted by liquid extraction and supercritical fluid extraction (SFE) and analyzed by GC/MS. The HS-GC/MS technique proved to be more useful for the volatiles than the liquid extraction GC/MS technique and the SFC method for less volatile substances. The extraction revealed that more than 140 substances were present, for example, aromatic substances such as limonene and aliphatic aldehydes from PET bottles that were previously used for fruit juice storage and aromatic aldehydes such as benzaldehyde.

Comparison of Bisphenol A (BPA) migration among others from infant bottles into infant formula was studied with the help of HPLC with fluorescence (FL) detectors [13]. As BPA was with native FL detectors no derivatization step is necessary. MS was used for confirmation. As the bottles were repeatedly used, the third migration solution was analyzed and the first and the second solutions were rejected. The test was conducted at 70 °C for 2 h followed by at 40 °C for 24 h. In acetic acid (3%), no migration of BPA was detected above the (limit of detection) LOD. Ethanol (50%) and infant formula sometimes showed low concentrations of BPA with the highest amount being 12 mg/kg food. However, GC/MS can also be used in the BPA detection in food after derivatization to improve detection.

Studies on properties of the polymer nanocomposite film after migration studies will give insights about the characteristics of the film during food storage. Polyhydroxybutyrate–cellulose nanocrystal (PHB-CNC)-based nanocomposite film showed improved thermal properties, high barrier for oxygen, flexibility and transparency [14]. OM studies carried out in polar and nonpolar solvents showed that the migration is within the limit. An after-migration characterization of the composite film revealed that there was a substantial decrease in the crystal formation and nucleation and a prominent decrease in the molecular weight was attributed to the migration of oligomeric lactic acid.

By using a combination of nanoparticles for preparing polymer composite films with improved properties, there is a risk of increased migration of ENPs. The incorporation of CNC and Ag nanoparticles combination into PLA matrix proved that the migration of silver ions was more prominent in the films where modified CNC was used compared to one without CNC. This was due to the high affinity of CNC, which caused the increased mobility of Ag ions leading to migration.

The effect of surface functionalized CNC [10] with phosphate ester of ethoxylated nonylphenol along with AgNPs in PLA films was studied. The inclusion of CNC and silver increased the barrier properties of PLA. The ternary system showed significant antibacterial effect and highest rate of disintegration under 15 days of composting conditions. OM test of the bionanocomposite was found below the permitted limits in both nonpolar and polar simulants, where it was found to be influenced by the CNC in the PLA matrix.

PLA composite containing AgNPs-based layered silicate as filler was tested for its biocidal properties against Gram-negative *Salmonella* spp [15]. The results of migration studies represent that AgNPs migration studies in aggressive food simulant (warm acetic acid), measured using voltammetry analysis, were found to be under the limit suggested by EFSA, even though there is an acceleration in migration rate after sixth day of exposure.

Successful deposition of hydrogenated amorphous carbon over PLA biofilms using the radiofrequency plasma-enhanced chemical vapor deposition technique found that the permeation of oxygen and water vapor through these PLA foils decreased [16]. The OM studies on the above hydrogenated amorphous carbon (a-C: H) films for the migration of nonvolatile substances in isooctane and ethanol were found to be below the legal standards. Higher migration rate in nonpolar simulant ethanol might be due to the thickness of a-C: H coating over PLA and some minor cracks were reported during the field emission scanning electron microscope analysis.

PLA nanocomposite film of modified clays, Clay1 (MMT + HDTA) and Clay2 (MMT + HDTA + ACO) with 4% clay content, for the food packaging application was extruded via twin screw extruder [15]. Migration and cytotoxicity studies were carried out on the above prepared film to estimate its toxicity limit on human exposure. The migrants from the aqueous food simulant water after OM were tested for its mutagenic effect in human intestinal Caco-2 and human liver HepG2 cells with an extract concentration of 2.5–100% and no mutagenic activity was found, showing a good profile for their use as FCMs.

Migration studies in real food matrices are carried out in case of testing the effect of active packaging in migrating the deliverables [15]. In this study migration of AgNPs from PVC was tested in meat of chicken and turkey for varying time and temperature. Inductively coupled plasma mass spectroscopy was used to quantify the amount of AgNPs in the food matrices after migration, which was found to be under the limit (0.03—8.4 mg/kg). The leaching of Ag ions may impose harm to the consumer and its leaching behavior severely relies on the time, temperature and concentration.

The use of layered double hydroxide (LDH) in melt-extruded PLA showed migration of oligomers of PLA with tin and laurate, which were used for the polymerization reaction [13]. The films were tested with 95% ethanol as a simulant for 10 days. All films showed migration of nano-LDH that were quantified by ICP-MS where the quantified migrants were under the limits.

9.4 Analytical techniques used for the characterization of migrants

9.4.1 Solid-phase microextraction (SPME)

Gas chromatography (GC) with MS is one of the common analytical instrument that can be used for the identification and quantification of the migrating species from a polymer matrix. Generally, GC is used to quantify lower molecular compounds with a straight forward approach. Moreover, additional steps are required for quantification in the case of aqueous simulants [16]. Trace amounts of sample from aqueous simulants can be extracted by using solid phase microextraction (SPME) attachment. A separation/extraction technique [16] uses fused silica fiber coated within a polymeric stationary phase in the liquid (immersion mode) or in liquid or solid phase. By using this method, the analytes in gas or liquid phase are adsorbed on to the silica fiber. Later the adsorbed analytes are desorbed by the system in the injection port. It is mainly used for the analysis of migrants from thermos oxidation and hydrolysis of polymeric product under test condition [16]. A detection limit of 5 μg/L of samples of crosslinked polyethylene pipes under aqueous food simulant is achieved using SPME to detect migrants such as styrene, phenol and benzophenone. SPME is tested with different fiber materials for adsorption of analytes; among those polydimethylsiloxane/divinylbenzene was efficient in all cases of analytes.

9.4.2 Electrospray ionization mass spectrometry (ESI-MS)

Nonvolatile migrants like UV stabilizers are commonly identified and quantified by liquid chromatography coupled to a mass spectrometric detector (LC-MS) [16]. In electrospray ionization (ESI), an ion source is often coupled with LC-MS, which is used for analyzing both nonvolatile and heat-sensitive compounds. Moreover, ESI-MS can be used independently without any chromatographic separation for the analysis of high molecular weight samples with rapid detection speed. Direct identification of migrant species is also possible with ESI-MS system as it is soft ionization technique where no fragmentation of the analyte occurs. In contrast, ESI in combination with MS-MS can also be utilized to identify fragments analytes. It can also detect degradation products of polyesters and oligomers of PET. No literature till date reports the use of ESI-MS system for detecting the additives migration [17]. In some cases, GC-MS separation technique is used to separate analytes before ESI-MS.

9.4.3 High-performance liquid chromatography (HPLC)

HPLC with UV detection is one of the robust technique used to identify and quantify high molecular weight polymeric compounds with precision. It is also used for quantification of high molecular additives like antioxidants, contain chromophore and so on. One such notable example was the use of HPLC in the analysis of polymer additives, where migration of cyclic oligomers from PET films to olive oil was observed [16], and in addition, HPLC-UV system is used to analyze stabilizers like Irgafos 168, antioxidants (Irganox 1076) from LDPE [16].

HPLC allows the separation of pigments, carbohydrates, vitamins, additives, mycotoxins, amino acids, proteins, lipids, chiral compounds and triglycerides in fats and oils [16]. Hydrodynamic chromatography (HDC) is also a very efficient technique to separate NPs in food simulant. HDC coupled with an ultraviolet–visible (UV-vis) detector has been used for the detection of colloidal size in suspensions and biomolecules from food samples [17]. Field–flow–fractionation (FFF), a technique similar to HPLC, can be used to separate NPs based on their thermal or hydraulic gradients, electrical forces and sedimentation rate [17]. It is a flexible elution technique where simultaneous separation and measurement can be done across a broad macromolecular colloidal particulate, ranging from about 1 nm to more than 100 nm [16]. A major advantage of this method is the lack of a stationary phase, which restrains the interaction between the sample and the equipment surfaces [16]. FFF can be used as an online or offline detection mode for the analysis of complex food samples [16]. FFF combined with HPLC and UV-vis light detectors are an efficient tool for the detection of NMs in food samples.

9.4.4 Spectroscopic and related techniques

The size, shape and structure of the nanomaterials in monodisperse system can be analyzed using small-angle X-ray scattering (SAXS), which uses X-ray as the principle technique for physical characterization in nanometer scale both in solid and liquid samples. Furthermore, SAXS is also used to analyze polydisperse systems where only particles size distribution can be calculated and studied. In the case of monodisperse systems, size, shape and structural determination of ENPs is possible. Other common techniques like UV-vis spectroscopy and Fourier transformation infrared (FTIR) are often used to characterize nanoparticles in solid and liquid sample [16]. Studies show that UV spectroscopy can be used to study the relationship between absorbance peak and particle size distributions in the case of quantum nanocrystals [15].

9.4.5 Nuclear magnetic resonance (NMR)

NMR is one of the potential techniques in elucidating the three-dimensional assembly of a particle in a solid or liquid sample. In addition, diffusion NMR spectroscopy has been proved to show the interaction of colloidal matter [16]. Pulsed field gradient NMR can be used to determine the coefficient of diffusion in fulvic acid.

Furthermore, matrix-assisted laser desorption/ionization is a mass spectrometry approach that can be used for the detection of nanoparticles that are fluorescently labeled. The development of fluorescence with MS approach seems to be a promising technique in the detection of nanoparticle in small quantity that can be analyzed in their original environment [16].

9.5 Regulations due to possible migration

Regulations are imposed on the usage of materials that are intended to come in contact with the food. There is no proper framework found for the usage of ENPs in the food packaging materials. The regulation made by each authority is based on their own local assessment. This exists because of the uncertain nature of the ENPs. Countries like New Zealand, the United States of America, Australia, European Union and Canada have passed revisions in FCMs that show general guidelines on nanoparticle usage in food packaging. In countries, such as China, Argentina, Brazil, Japan and Mexico no such visibility is provided. Other than the guidelines, none of the countries made amendments to the directives [8].

European Union regulations before 2004 limited the use of NPs in FCMs. In regulation no. 1935/2004 amendments for the first-time European Parliament made a note about the use of "active and intelligent materials that come into contact with food." Even though no specific mention about the ENPs was made, a specific mention of the terms "active" and "intelligent" materials was registered. Other EC regulation no. 10/2011 (European Commission (EU), 2011) and 82/711/EEC (European Commission (EU), 1982) and 85/572/EEC detail about the procedure for overall and SM studies and its related food simulants.

The migration limit set by the EU regulation no. 10/2011 for all FCMs cannot exceed a maximum limit of 10 mg/dm^2 from package to food. Risk assessment strategy is followed to set the limit for SM of a substance depending on the amount of risk it can impose. The status of nanoparticles in FCMs regulation is outlined in Figure 9.1.

The United States Food and Drug Administration (USFDA) follows a strict rule for the use of nanomaterials in FCMs inside United States of America. In a recent update by USFDA, the use of FCMs requires a certificate from the manufacturer regarding the toxicity status of the material in human and animal under worst case scenario (United States Food and Drugs Association [USFDA], 2014).

In the regulation, it is stated that usage or reduction of materials into nanoscale in FCM is considered to be deviating from the regulation and has to provide a proper documentation about the usage of such material including the toxicity assessment by the product manufacturer. USFDA implies that "it does not make generalizations related to the harmfulness of all products containing nanotechnology but instead encourages industry to provide specific risk assessment so that products can be assessed on a case-by-case basis."

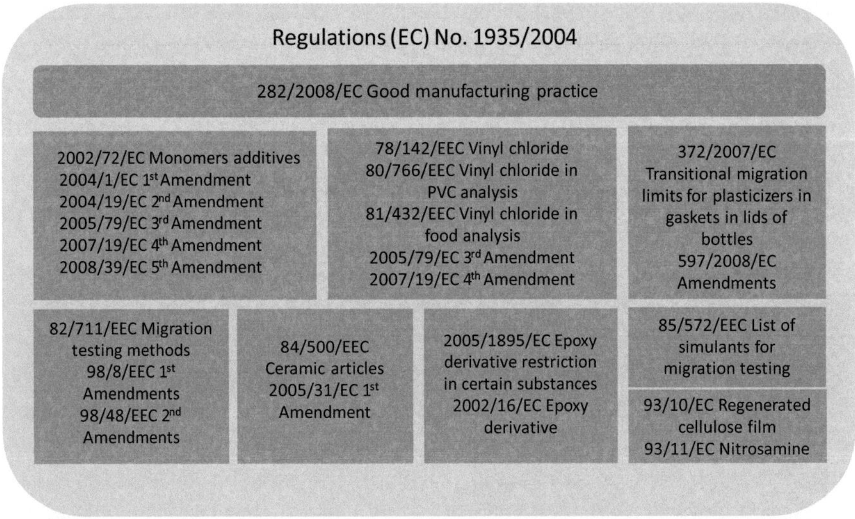

Figure 9.1: EU regulation on FCM.

Talking in terms of nanomaterials in food packaging, an important argument is that nanomaterials are more prone to agglomeration [8]. Even though many problems occur during processing, immobilization of the nanomaterials after processing showed a reduced risk of migration. In contrast, this immobilization can impart problem in achieving the required properties.

9.6 Conclusion

Food encounters a variety of materials from the process of production and processing, in packing condition and storage until it is consumed. Contamination is caused by the migrants from the FCMs to the food that have a tendency to change the organoleptic properties of the food. More amendments are required in the legislative directives of various countries on the usage of ENPs in FCMs. Research on polymer nanocomposite combined with migration studies can give visibility to the industries and consumers about the risk associated with nanoparticles. An international body of authority can be made to govern and provide easy policies in using this rapidly growing nanomaterials.

References

[1] López-Cervantes, J., Sánchez-Machado, DI., Simal-Lozano, J., & Paseiro-Losada, P. Migration of ethylene terephthalate oligomers from roasting bags into olive oil. Chromatographia 2003, 58, 321–326.

[2] Maisanaba, S., Pichardo, S., Jordá-Beneyto, M., Aucejo, S., Cameán, AM., & Jos, Á. Cytotoxicity and mutagenicity studies on migration extracts from nanocomposites with potential use in food packaging. Food Chem Toxicol 2014, 66, 366–372.

[3] Mattioli, S., Peltzer, M., Fortunati, E., Armentano, I., Jiménez, A., & Kenny, JM. Structure, gas-barrier properties and overall migration of poly (lactic acid) films coated with hydrogenated amorphous carbon layers. Carbon 2013, 63, 274–282.

[4] Santillana, MI., Ruiz, E., Nieto, MT., Bustos, J., Maia, J., Sendón, R., & Sanchez, JJ. Migration of bisphenol A from polycarbonate baby bottles purchased in the Spanish market by liquid chromatography and fluorescence detection. Food Addit Contam Part A 2011, 28, 1610–1618.

[5] Ulery, BD., Nair, LS., & Laurencin, CT. Biomedical applications of biodegradable polymers. J Polym Sci B 2011, 49, 832–864.

[6] Yang, Y., Doudrick, K., Bi, X., Hristovski, K., Herckes, P., Westerhoff, P., & Kaegi, R. Characterization of food-grade titanium dioxide: the presence of nanosized particles. Environ Sci Technol 2014, 48, 6391–6400.

[7] Shirai, MA., Grossmann, MVE., Mali, S., Yamashita, F., Garcia, PS., & Müller, CMO. Development of biodegradable flexible films of starch and poly (lactic acid) plasticized with adipate or citrate esters. Carbohydr Polym 2013, 92, 19–22.

[8] Othman, SH. Bio-nanocomposite materials for food packaging applications: types of biopolymer and nano-sized filler. Agric Agric Sci Procedia 2014, 2, 296–03.

[9] Kerry, J., & Butler, P. Legislative issues relating to smart packaging, Kerry J, Butler P., Eds, Smart packaging technologies for fast moving consumer goods, John Wiley & Sons, Chichester, 2008, 305–320.

[10] Ahvenainen R., ed. Novel food packaging techniques. Elsevier, Woodhead publishing Limited and CRC Press LLC, Boca Raton, Florida, 2003.

[11] Prasad, P., & Kochhar, A. Active packaging in food industry: a review. J Environ Sci Toxicol Food Technol 2014, 8, 1–7.

[12] Zhao, R., Torley, P., & Halley, PJ. Emerging biodegradable materials: starch-and protein-based bio-nanocomposites. J Mater Sci 2008, 43, 3058–3071.

[13] Schmidt, B., Katiyar, V., Plackett, D., Larsen, EH., Gerds, N., Koch, CB., & Petersen, JH. Migration of nanosized layered double hydroxide platelets from polylactide nanocomposite films. Food Addit Contam A 2011, 28, 956–966.

[14] Dhar, P., Bhardwaj, U., Kumar, A., & Katiyar, V. Poly (3-hydroxybutyrate)/cellulose nanocrystal films for food packaging applications: barrier and migration studies. Polym Eng Sci 2015, 55, 2388–2395.

[15] Ke, T., & Sun, XS. Starch, poly (lactic acid), and poly (vinyl alcohol) blends. J Polym Environ 2003, 11, 7–14.

[16] Brockgreitens, J., & Abbas, A. Responsive food packaging: recent progress and technological prospects. Compr Rev Food Sci F 2016, 15, 3–15.

[17] Yam KL., ed. The Wiley encyclopedia of packaging technology. John Wiley & Sons, Inc., United States, 2009.

Tabli Ghosh and Vimal Katiyar

10 Edible polymer-based sustainable food packaging

10.1 Introduction

Inclusion of several postharvest techniques in combination may lessen the natural deterioration of the food commodities, whereas packaging can contribute toward the protection of food products from surroundings [1]. Postharvest techniques including packaging can reduce the postharvest diseases, respiration rate, senescence and surface microbial count of fruits and vegetables. One such available postharvest technique is the addition of thin layer of edible materials on food surface, which in general is known as *edible coating* [2]. Edible coating acts as a semi-permeable barrier against surroundings, which protects the produce from oxygen, water vapor, fungi and other harmful environmental agents [3]. It enhances the shelf life of food products without creating any harmful effects to them, and in addition, it has many advantages over other postharvest techniques such as heat treatment, irradiation, use of chemicals and so on. Edible coating includes the methods of dip coating and spray coating, which helps to create a barrier between environment and food products, and also increases the aesthetic property of the food by giving a glossy appearance to the surface of the food [4]. In the early ages, consumers being unaware about the significance of edible coatings were unwilling to accept the coated food. However, the coated food gained acceptance after people became familiar with the significance of edible coated food products. Application of edible coating helps to increase the shelf life of food by reducing microbial growth and mold growth on the food surface. Besides, application of edible coating with additional functional ingredients to food is also considered as another effective way to enhance the shelf life and quality of food products.

Furthermore, the essence of innovation of biopolymers can overcome the problem of waste disposal in the environment [5]. Since decades, cellulosic biopolymer-based materials are being used as secondary packaging, which creates a significant contribution to the environment due to their ecofriendly nature. However, biopolymers are affected by relative humidity, temperature, microorganism spoilage and many other factors [5] unlike fossil-based polymers [6]. In today's world, the use of biopolymers is another wide-spreading part of food-processing technology [7]. However, its limitations lie within its durability and shelf life of the material [8]. These limitations outburst with the concept of using edible coating for food-packaging applications.

Further, the combination of edible coating and biopolymers is a promising technique to preserve and protect food products from surroundings and act as a

https://doi.org/10.1515/9783110648034-010

safeguard against mechanical/thermal/chemical injuries. Although edible coating has good barrier properties against surroundings, a single layer of edible coating cannot serve as a productive barrier. So biopolymers and edible coating together can meet the shortcomings of each other [5].

10.2 Edible coating

10.2.1 Methods of coating

Based on the process of application of coating material on the surface of food products, there are generally four types of methods available: (1) dipping, (2) spraying, (3) foaming and (4) brushing. Among available methods, dip coating and spray-coating methods are extensively used throughout the world. Both the methods involve some stages to give a proper coating on the surface of the food product and the quality of coating material is examined based on various property analysis (Figure 10.1).

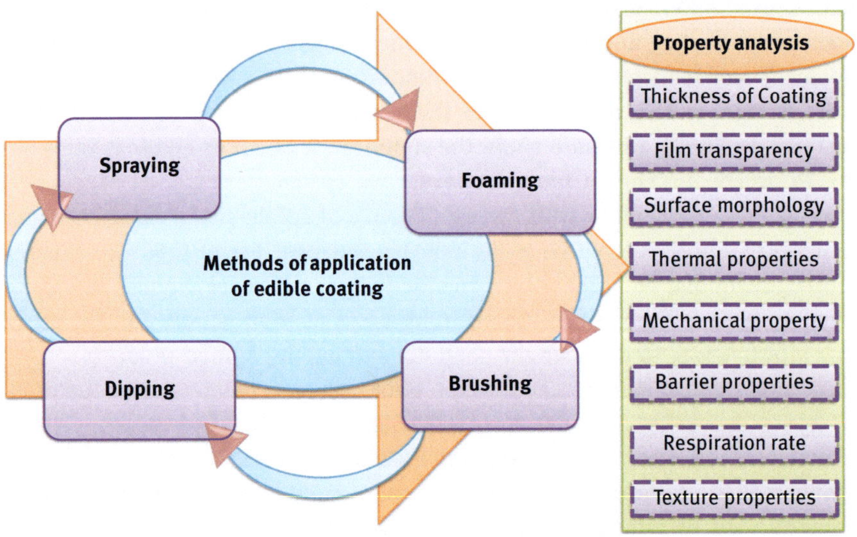

Figure 10.1: Methods of application of edible coating material on food product.

Dip coating is a method of edible coating, where the food product is dipped in the prepared coating solution, thereby developing a coat on the food surface. The disadvantages of using this method include increasing microbial load in coating solution due to multiple coating cycles, which may get transferred to other food products too. Moreover, dip coating often provides nonuniform coat thickness to the food products. To overcome the limitations of this method, food surface may be treated with some antimicrobial solution for reducing microbial count on the surface of food and probes can also be used during dipping for holding the sample, which reduces hand contact. On the other hand, in spray coating, coating materials are sprayed on the food surface using nozzles and the approaches of spray-coating methods commonly include electrospinning [9] and supercritical fluid spraying [10]. Further, spray-coating methods attract more concentration for giving edible coating to the food product that does not come in contact with food, thus reducing waste of edible materials.

10.2.2 Types of coating materials

Coating materials are mainly classified based on several materials utilized for coating the food products. Table 10.1 gives details about the materials used in edible coating of food product. Different coating materials have various advantages and disadvantages; thus, combinations of coating materials are used to overcome the limitations of each other. Polysaccharides-based derivatives, being hydrophilic in nature, have poor barrier properties against water. However, using a lipid derivative with polysaccharides overcomes the problem. Based on the requirement to retain the property of food, coating materials can be used alone or in combination with others.

Polysaccharide-based edible coating can form a supporting matrix structure, which can be fixed and inflexible gel. Moreover, it has an intermediate permeability toward gas-exchange properties. However, hydrophilic nature of polysaccharides makes them poor water barrier coatings, restricting their usage over other coating materials [11]. Subtypes of polysaccharide-based coating are listed in Table 10.1. One of the polysaccharide biopolymers is chitin, which is a compound of *N*-acetyl-D-glucosamine in invertebrates and insects, which can be converted to chitosan, another polysaccharide derivative, in the presence of concentrated alkali by algae [12].

Table 10.1: Types and subtypes of coating materials.

Types of coating materials	Subtypes of coating materials
Polysaccharides	Cellulose, starch, chitosan, pectin, carrageenans, alginate, xanthan, gums
Protein	Corn zein, wheat gluten, whey protein, casein, soy protein
Lipid and wax	Vegetable oil, mineral oil, olive oil, oleic acid, acetoglycerides, surfactants, carnauba wax, bees wax, petroleum-based waxes, rice bran waxes
Composites	Cellulose with bees wax, alginate with casein

Chitosan is derived from natural sources. It is nonallergic, biocompatible polymer and has many health beneficial properties [13]. The use of chitosan on strawberry helps to increase the shelf life of the product by reducing respiration and delaying ripening. Edible coating based on pectin extracted from apple is applied on fresh-cut persimmon for browning inhibition and microbial control [14].

Protein-based coating materials are hydrophilic in nature, which makes them poor water-repellent coating materials [15]. The combined use of protein and wax material can overcome the shortcomings of protein coating. Protein-based coatings are mainly developed from corn zein, whey protein, wheat gluten, peanuts and many others [16].

Lipid and wax materials have better water retention capacity over polysaccharide and protein coating materials [17]. Lipid in the form of oil sacs can be used on food surface. The presence of lipid-based coating between environment and food makes a barrier for water permeation, which causes negligible shrinkage in the food during storage. Lipid and waxes comprise of vegetable oil, mineral oil, acetylglycerides, oleic acid, beeswax, rice bran waxes and so on [18, 19]. Waxes give a glazy appearance to the food for aesthetic purposes. The limitation of this coating material is its poor gas permeability, which can be balanced by using other materials such as polysaccharides and proteins in combination.

Combination of coating materials can be developed by combining starch with protein, lipid with starch, starch with wax, protein with starch, lipid with wax, wax with protein and so on. One such example reported is the combined use of hydroxypropyl methylcellulose with beeswax on cold stored cherry tomatoes [20].

10.3 Active agents in edible coating

Edible coating has a notable application in providing protection and functionality to the food [21]. The addition of antioxidant agents [22], antidiabetic agents [23], antimicrobial agents [24], antifungal agents [4] and antihypertensive agents [25] to the food product leads to an increase in functionality and nutritional value of the food product. Most commonly used extraction techniques for extracting such active agents are solvent extraction, microwave extraction, ultrasonic extraction and so on. The active agents can easily be extracted from natural available fruits and vegetables, spices, medicinal plants and other natural sources. With many medicinal plants available around the world, the idea of adding functionality to produce from natural available sources outburst with new prospects of using edible coating. Table. 10.2 describes the beneficial outcome of application of edible coating on food products.

Table 10.2: Application of edible coating with their beneficial outcome.

Food product	Coating materials	Discussion	References
Strawberry	Starch, carrageenan, chitosan	Retard senescence	[26]
Cut apple and potato	Cellulose, soy protein	Improve storage life	[27]
Tomato	Gum arabic	Improve shelf life and postharvest life	[28]
Sliced mango	Chitosan	Improve quality and shelf life	[29]
Table grapes	Aloe vera gel	Improve quality and safety	[30]
Vacuum fried banana chips	Xanthan gum, guar gum	Improve quality, safety	[31]
Banana	Gum arabic and chitosan	Reduce anthracnose disease	[32]
Fresh cut papaya	Alginate and gellan	Improve texture and nutrition	[33]
Guava	Chitosan	Control weight, delayed ripening	[34]
Fresh cut broccoli	Chitosan	Provide safety	[35]
Carrot sticks	Chitosan and moderate modified atmosphere	Improve quality and provide phytochemicals	[36]
Apple, potato	Milk protein	Prevent oxidative browning	[37]
Fried food products	Methylcellulose	Reduce oil uptake	[38]

10.3.1 Carriers of antioxidants

Incorporation of antioxidants to the coating material leads to the increased health benefits of food. Antioxidant agents have a property of scavenging free radicals generated from the body and reduce oxidative stress with no adverse effect. On the other hand, chemical preservatives have the disadvantages of their toxin-producing capability above the permissible limit. Naturally used antioxidants are extracted from fruits and vegetables, spices, crops such as green tea polyphenols, fig (*Ficus carica*), ginger polyphenols and so on. The study of combined use of rice starch-based edible coating with antioxidants from tea leaf extract on tomato is reported to have better efficacy toward health benefits [2].

10.3.2 Carriers of antidiabetic agents

Incorporation of antidiabetic agents to the coating material increases the health benefits of diabetic patient. Coating materials with optimized amount of extracts of *karela, Terminilia chebula*, radish, pomegranate ginger, fenugreek seeds, cinnamon and coriander provides antidiabetic properties to food materials.

10.3.3 Carriers of anticancer agents

Many a times, edible coating acts as a carrier material for anticancer agents, which has greater benefits to the mankind. Many fruits and vegetables contain anticancer agents as bioactive compounds, which can be incorporated in the coating material to add functionality to the food. Anticancer agents can be extracted from fig (*Ficus carica*), green tea extract, seeds of bitter apricot, chiraytata, tomato, raspberries, strawberries and so on.

10.3.4 Flavor enhancer

The addition of flavoring agents to the coating materials leads to the enhanced acceptability of the coated food [37]. The use of ginger, cardamom, cinnamon and other spices provides their characteristic flavors to coating materials.

10.4 Application of edible coating

Most of the fruits and vegetables have a very short postharvest life. Application of edible coating reduces respiration rate, attack by insects, microorganism, mechanical injuries and thermal injuries, providing better shelf life to perishable food products. Edible coating is mainly applied on meat and meat products, fruits and vegetables, cut fruits, minimally processed foods, fried food products and so on.

10.5 Properties of coating materials

The developed coating material is tested for its quality based on various property analyses such as surface property, film transparency, film thickness, color kinetics, weight loss, titrable acidity, pH, respiration kinetics of fruit and so on. Surface morphology of the coating is generally determined by using microscopic instrument "scanning electron microscopy" [38]. Film transparency detects the light transparency of the coating [2]. Larger coating particles act as barriers for light transparency [2]. Thickness of the coating materials is measured using micrometer. Another most important factor is weight loss, which has a direct correlation with the coating property, and is measured using laboratory weighing balance. The coating material having better retaining qualities against various environmental factors leads to reduced weight loss of food products with storage time period. In addition, the decrease in respiration rate of the food defines the good coating behavior. The coat reduces the respiration rate of commodities and is measured by using gas analyzer, which is also known as O_2/CO_2 analyzer [39]. Acidity and pH of tomatoes during storage depict the barrier properties of coating materials. One of the crucial factors is microbial count of the coated food during storage, which depicts its transportation conditions. Another crucial factor is the texture of the coating materials as the consumer acceptance of the coated food is determined using texture analyzer.

10.6 Case study

10.6.1 Effect of edible coating on papaya fruits

Narsaiah et al. studied the effect of bacteriocin (antimicrobial agent) incorporating alginate-based edible coating on the postharvest quality of minimally processed papaya (*Carica papaya L.*) fruits [40]. The coating material alginate acts as a delivery material for antimicrobial agents (bacteriocin), which helps in prolonging shelf life of papaya fruits by decreasing microbial load on the fruits. For this study, development of edible coating was done by incorporating specified amount of bacteriocins (0, 20% w/v) in alginate (1, 1.5, 2% w/v) solution. Application of developed coating materials on papaya fruits resulted in decreasing O_2 concentration approximately by 1.8 times than that in uncoated sample. Furthermore, acidity of the edible coated sample was observed to be lessened by 2.7 times than uncoated sample. An increase in acidity leads to the development of toxic compounds to the food products, causing food intoxication. The use of antimicrobial agents helps in decreasing the microbial load on the selected samples. Application of coating materials on minimally processed papaya sample was found to decrease the microbial load of sample approximately by 10^5 times than uncoated fruit sample. Edible coating acts as a barrier against oxygen transfer rate, maintain firmness, decrease toxic effects, decline senescence and decrease respiration rate of minimally processed papaya fruits. Finally, the study depicts that 2% alginate-based coating was more suitable than alginate solution (1, 1.5%) for a storage study of 21 days.

10.6.2 Effect of edible coating on mango fruits

Chien et al. studied the chitosan-based edible coating on sliced mango fruits [27]. Mango is one of most eatable fruit product. Mango slices were treated with chitosan solution (0, 0.5, 1 and 2%) at storage temperature of 6 °C. The chitosan-coated sliced mangoes were kept in plastic trays and were overwrapped with polyvinylidene chloride film. The observed data showed that chitosan (2%) coated sample and uncoated control samples have a weight loss of 10.3% and 19.9%, respectively. Besides, application of chitosan solution (2%) resulted in a decreased microbial load of sliced mango. The microbial load of control and edible coated mango slices was found to be in the ranges of 3.82–6.41 log CFU/g and 3.82–5.41 log CFU/g, respectively. This depicts that chitosan has a potential to provide better postharvest quality of sliced mangoes.

10.6.3 Effect of edible coating on tomato fruits

Ali et al. studied the effect of applying gum arabic (5, 10, 15 and 20%) on tomato (20 °C and 80–90% RH) for 20 days storage period [41]. This study showed that gum arabic has a considerable property retaining capacity for carotenoid content, phenolic content and antioxidant capacity with an increase in shelf life.

10.6.4 Effect of edible coating on strawberry fruits

Application of chitosan (0.5, 1 and 1.5%) on strawberry fruits was found to affect its antioxidant activity, flavonoid content and shelf life of the fruit at various storage temperatures (5 and 10 °C) [22]. The study reveals that chitosan-coated (0.5%) strawberry fruits have less percentage of decay (11.7%) than uncoated strawberry sample (21.5%) at 5 °C. Again, at 10 °C, the observed percentage decay was found to be 25.3, 15.7 6.9% for 0.5, 1, 1.5% chitosan-coated samples, respectively, and for control samples at 10 °C the percentage decay was 95.2%. This depicts the significance of chitosan-coating materials on strawberry fruit products.

10.7 Conclusion

Research findings surpassingly benefit human society through innovations of edible coatings for food packaging applications. Conventional plastic materials have versatile applications for their use in food packaging but the prevailing plastic materials being nondegradable lead to the development of environmental-friendly plastic materials, which can reduce the biochemical changes occurring in food during storage and transportation, with no adverse effects on the toxicity content and aesthetics property of the food. However, biopolymers have certain drawbacks such as low thermal resistance and poor mechanical properties, which need to be tailored for improved shelf life. Edible coating is a semipermeable barrier of edible materials (polysaccharides, protein, lipid and composites) on food surface having properties of delivering functionality to the food and it overcomes the limitations of biopolymers. The combined use of edible coating and biopolymers in food packaging significantly enhance the shelf life of food products along with limiting transportation injuries. Thus edible coating with biopolymers is considered to have great potential in food packaging applications.

References

[1] Robertson, GL. Food packaging: principles and practice, CRC press, Boca Raton, London, New York, 2016.

[2] Das, DK., Dutta, H., & Mahanta, CL. Development of a rice starch-based coating with antioxidant and microbe-barrier properties and study of its effect on tomatoes stored at room temperature. LWT-Food Sci Technol 2013, 50, 272–278.

[3] Azarakhsh, N., Osman, A., Ghazali, HM., Tan, CP., & Adzahan, NM. Lemongrass essential oil incorporated into alginate-based edible coating for shelf-life extension and quality retention of fresh-cut pineapple. Postharvest Biol Technol 2014, 88, 1–7.

[4] Vásconez, MB., Flores, SK., Campos, CA., Alvarado, J., & Gerschenson, LN. Antimicrobial activity and physical properties of chitosan–Tapioca starch based edible films and coatings. Food Res Int 2009, 42, 762–769.

[5] Huang, SJ. Polymer waste management–Biodegradation, incineration, and recycling. J Macromol Sci A 1995, 32, 593–597.

[6] Harding, KG., Dennis, JS., Von Blottnitz, H., & Harrison, STL. Environmental analysis of plastic production processes: comparing petroleum-based polypropylene and polyethylene with biologically-based poly-β-hydroxybutyric acid using life cycle analysis. J Biotechnol 2007, 130, 57–66.

[7] Peelman, N., Ragaert, P., De Meulenaer, B., Adons, D., Peeters, R., Cardon, L., & Devlieghere, F. Application of bioplastics for food packaging. Trends Food Sci Technol 2013, 32, 128–141.

[8] Galgano, F. Biodegradable packaging and edible coating for fresh-cut fruits and vegetables. Ital J Food Sci 2015, 27, 1–20.

[9] Ganesh, VA., Nair, AS., Raut, HK., Walsh, TM., & Ramakrishna, S. Photocatalytic superhydrophilic TiO 2 coating on glass by electrospinning. Rsc Adv 2012, 2, 2067–2072.

[10] Lee, C., Hoy, KL., & Donohue, MD. inventor; Union carbide chemicals and plastics company Inc. assignee 1990, US4923720.

[11] Oms-Oliu, G., Soliva-Fortuny, R., & Martín-Belloso, O. Using polysaccharide-based edible coatings to enhance quality and antioxidant properties of fresh-cut melon. LWT-Food Sci Technol 2008, 41, 1862–1870.

[12] Brasil, IM., Gomes, C., Puerta-Gomez, A., Castell-Perez, ME., & Moreira, RG. Polysaccharide-based multilayered antimicrobial edible coating enhances quality of fresh-cut papaya. LWT-Food Sci Technol 2012, 47, 39–45.

[13] Suh, JKF., & Matthew, HW. Application of chitosan-based polysaccharide biomaterials in cartilage tissue engineering: a review. Biomaterials 2000, 21, 2589–2598.
[14] Sanchís, E., González, S., Ghidelli, C., Sheth, CC., Mateos, M., Palou, L., & Pérez-Gago, MB. Browning inhibition and microbial control in fresh-cut persimmon (Diospyros kaki Thunb. cv. Rojo Brillante) by apple pectin-based edible coatings. Postharvest Biol Technol 2016, 112, 186–193.
[15] Mchugh, TH., Weller, CL., & Krochta, JM. Edible coatings and films to improve food quality, CRC press, Boca Raton, Florida, 1994, 201.
[16] Seydim, AC., & Sarikus, G. Antimicrobial activity of whey protein based edible films incorporated with oregano, rosemary and garlic essential oils. Food Res Int 2006, 39, 639–644.
[17] Garcia, MA., Martino, MN., & Zaritzky, NE. Lipid addition to improve barrier properties of edible starch-based films and coatings. J Food Sci 2000, 65, 941–944.
[18] Lin, D., & Zhao, Y. Innovations in the development and application of edible coatings for fresh and minimally processed fruits and vegetables. Compr Rev Food Sci Food Saf 2007, 6, 60–75.
[19] Baldwin, EA., Nisperos-Carriedo, MO., & Baker, RA. Use of edible coatings to preserve quality of lightly (and slightly) processed products. Crit Rev Food Sci Nutr 1995, 35, 509–524.
[20] Fagundes, C., Palou, L., Monteiro, AR., & Pérez-Gago, MB. Effect of antifungal hydroxypropyl methylcellulose-beeswax edible coatings on gray mold development and quality attributes of cold-stored cherry tomato fruit. Postharvest Biol Technol 2014, 92, 1–8.
[21] Silva-Weiss, A., Ihl, M., Sobral, PJA., Gómez-Guillén, MC., & Bifani, V. Natural additives in bioactive edible films and coatings: functionality and applications in foods. Food Eng Rev 2013, 5, 200–216.
[22] Wang, SY., & Gao, H. Effect of chitosan-based edible coating on antioxidants, antioxidant enzyme system, and postharvest fruit quality of strawberries (Fragaria x aranassa Duch.). LWT-Food Sci Technol 2013, 52, 71–79.
[23] Chinchilla, N., Carrera, C., Durán, AG., Macías, M., Torres, A., & Macías, F. A. Aloe barbadensis: how a miraculous plant becomes reality. Phytochem Rev 2013, 12, 581–602.
[24] Rojas-Graü, MA., Raybaudi-Massilia, RM., Soliva-Fortuny, RC., Avena-Bustillos, RJ., McHugh, TH., & Martín-Belloso, O. Apple puree-alginate edible coating as carrier of antimicrobial agents to prolong shelf-life of fresh-cut apples. Postharvest Biol Technol 2007, 45, 254–264.
[25] Dangaran, K., Tomasula, PM., & Qi, P. Structure and function of protein-based edible films and coatings, Edible films and coatings for food applications, Springer, New York, NY, 2009, 25–56.
[26] Ali, A., Maqbool, M., Ramachandran, S., & Alderson, PG. Gum arabic as a novel edible coating for enhancing shelf-life and improving postharvest quality of tomato (Solanum lycopersicum L.) fruit. Postharvest Biol Technol 2010, 58, 42–47.
[27] Chien, PJ., Sheu, F., & Yang, FH. Effects of edible chitosan coating on quality and shelf life of sliced mango fruit. J Food Eng 2007, 78, 225–229.
[28] Valverde, JM., Valero, D., Martínez-Romero, D., Guillén, F., Castillo, S., & Serrano, M. Novel edible coating based on Aloe vera gel to maintain table grape quality and safety. J Agric Food Chem 2005, 53, 7807–7813.
[29] Sothornvit, R. Edible coating and post-frying centrifuge step effect on quality of vacuum-fried banana chips. J Food Eng 2011, 107, 319–325.
[30] Maqbool, M., Ali, A., Ramachandran, S., Smith, DR., & Alderson, PG. Control of postharvest anthracnose of banana using a new edible composite coating. Crop Prot 2010, 29, 1136–1141.

[31] Tapia, MS., Rojas-Graü, MA., Carmona, A., Rodríguez, FJ., Soliva-Fortuny, R., & Martin-Belloso, O. Use of alginate-and gellan-based coatings for improving barrier, texture and nutritional properties of fresh-cut papaya. Food Hydrocolloid 2008, 22, 1493–1503.
[32] Hong, K., Xie, J., Zhang, L., Sun, D., & Gong, D. Effects of chitosan coating on postharvest life and quality of guava (Psidium guajava L.) fruit during cold storage. Sci Hort 2012, 144, 172–178.
[33] Alvarez, MV., Ponce, AG., & Moreira, MDR. Antimicrobial efficiency of chitosan coating enriched with bioactive compounds to improve the safety of fresh cut broccoli. LWT-Food Sci Technol 2013, 50, 78–87.
[34] Simões, AD., Tudela, JA., Allende, A., Puschmann, R., & Gil, MI. Edible coatings containing chitosan and moderate modified atmospheres maintain quality and enhance phytochemicals of carrot sticks. Postharvest Biol Technol 2009, 51, 364–370.
[35] Le Tien, C., Vachon, C., Mateescu, MA., & Lacroix, M. Milk protein coatings prevent oxidative browning of apples and potatoes. J Food Sci 2001, 66, 512–516.
[36] Garcia, MA., Ferrero, C., Campana, A., Bértola, N., Martino, M., & Zaritzky, N. Methylcellulose coatings applied to reduce oil uptake in fried products. Food Sci Technol Int 2004, 10, 339–346.
[37] Martín-Belloso, O., Rojas-Graü, MA., & Soliva-Fortuny, R. In Edible films and coatings for food applications, Eds., M. E. Embuscado and K.C. Huber, Springer, New York, 2009, 295.
[38] García, MA., Martino, MN., & Zaritzky, NE. Edible starch films and coatings characterization: scanning electron microscopy, water vapor, and gas permeabilities. Scanning 1999, 21, 348–353.
[39] Sutcliffe, DW., Carrick, TR., & Moore, WH. An automatic respirometer for determining oxygen uptake in crayfish (Austropotamobius pallipes (Lereboullet)) over periods of 3–4 days. J Exp Biol 1975, 63, 673–688.
[40] Narsaiah, K., Wilson, RA., Gokul, K., Mandge, HM., Jha, SN., Bhadwal, S., & Vij, S. Effect of bacteriocin-incorporated alginate coating on shelf-life of minimally processed papaya (Carica papaya L.), Postharvest Biol Technol, 2015, Vol. 100, 212–218.
[41] Ali, A., Maqbool, M., Alderson, PG., & Zahid, N. Effect of gum arabic as an edible coating on antioxidant capacity of tomato (Solanum lycopersicum L.) fruit during storage. Postharvest Biol Technol 2013, 76, 119–124.

Naba Kumar Kalita, Melakuu Tesfaye, Purabi Bhagabati
and Vimal Katiyar

11 Trends on end-of-life options, including recycling, reusing and composting of waste food packages

11.1 Introduction

Recyclability is one of the important criteria of sustainable technology of carbon sources in the case of fossil as well as biobased products. Recycling is a process that is cost effective, and recycling of bioplastics will also have serious implications on the future of our environment. Plastic-based environmental pollution has become a major concern of today's world due to its large consumption and poor disposal techniques [1]. Day-to-day activities of human life are simplified by the products made from plastics including toothpaste, drinking bottles, food packaging, sports equipment, cars, computers electrical wires (insulators), pipes and so on [2]. In 2013 global plastic production was estimated to be more than 250 MT (metric ton), of which around 40% accounted for packaging. According to the studies, most of plastics used for food and other simple product packaging are utilized only for 12 min, which are then disposed in to the environment to be there for more than 400 years. From the total world consumption of plastics around 43% are disposed in landfills and more than 10 million tons end up in the ocean each year [3]. Waste recovery using different recycling techniques has been developed to minimize the waste disposal and convert into some valuable form [4, 5]. However, compared to the yearly plastic production growth rate and plastics that are already deposited on the earth for the last 50 years, world's recycling efficiency is not sufficient enough to tackle the problem. Additionally, some of the recycling techniques like energy recovery from the plastic waste by direct incineration create significant air pollution [2]. Moreover, commodity plastics are produced from the petroleum resource, which will soon be depleted posing a serious concern on world energy security.

Biobased polymers have a promising performance to substitute petroleum-based plastics. Research in the field of bioplastics can potentially be beneficial in increasing the efficiency of recycling waste management and reducing the greenhouse emissions [6]. The natural decomposition process of these synthetic biopolymers into nontoxic components (CO_2 and CO) makes them suitable over petroleum-based polymers [7, 8]. However, their use is limited due to their high cost. To replace conventional polymers, the biopolymers need to be cost effective for large-scale production [9]. One way of minimizing the cost is by using

https://doi.org/10.1515/9783110648034-011

the technologies that can help to minimize the energy during the synthesis and processing of the material. Recycling has been known to be the efficient method toward energy minimization and releasing less carbon dioxide than producing a new product. By doing so, it is possible to reduce the extensive utilization of feedstock with low cost of the material. Therefore, appropriate recycling mechanism needs to be developed to ensure ultimate utilization of bioplastics. Based on this fact, allowing the bioplastics to degrade naturally into carbon dioxide and water can be considered as wasting the energy spent on the production process. Currently, bioplastics are produced in very small quantities, holding 2% shear of polymer production in 2013, which is expected to grow up to 4% by 2020 [10]. In addition to this, the technology for recycling has not yet fully been developed.

This chapter will elaborate the recycling technologies of biodegradable plastics including mechanical recycling, chemical/enzymatic recycling, pyrolysis and incineration and composting. Case studies will categorically be included in terms of PLA, which is the most reported biobased plastic till date. However, biodegradable plastics like polyhydroxyalkanoates (PHA), poly(butylene-succinate) (PBS) and starch-based biodegradable plastics that are used as base matrix and their blends and composites recycling will also be discussed [11–16]. Finally, the challenges and future opportunities of recycling biopolymers will be described.

11.2 Recycling of biodegradable plastics

Recycling of biodegradable plastics is becoming an emerging area of research, with the objective of finding optimum utilization of the material instead of natural degradation. Figure 11.1 illustrates a schematic representation of recycling of biodegradable plastic material, which has positive impact not only on the environment but also to the economy.

Hence, recyclability of biodegradable polymers can be an effective criterion for the benefit of the environment. Generally, the recycling strategies, which are reported in literature, can be categorized into reprocessing, chemical recycling, pyrolysis and composting. According to the comparative study done by Piemonte and coworkers, reprocessing is more preferable than other waste utilization techniques such as composting and pyrolysis/incineration in terms of feedstock consumption, energy saving and greenhouse gases emission. Effective recycling of biodegradable plastics (e.g., PLA) into its monomer or lower molecular weight re-polymerizable oligomer is also getting increasing attention in literature since low-cost monomeric unit can be generated when compared to the usual fermentation processes [17].

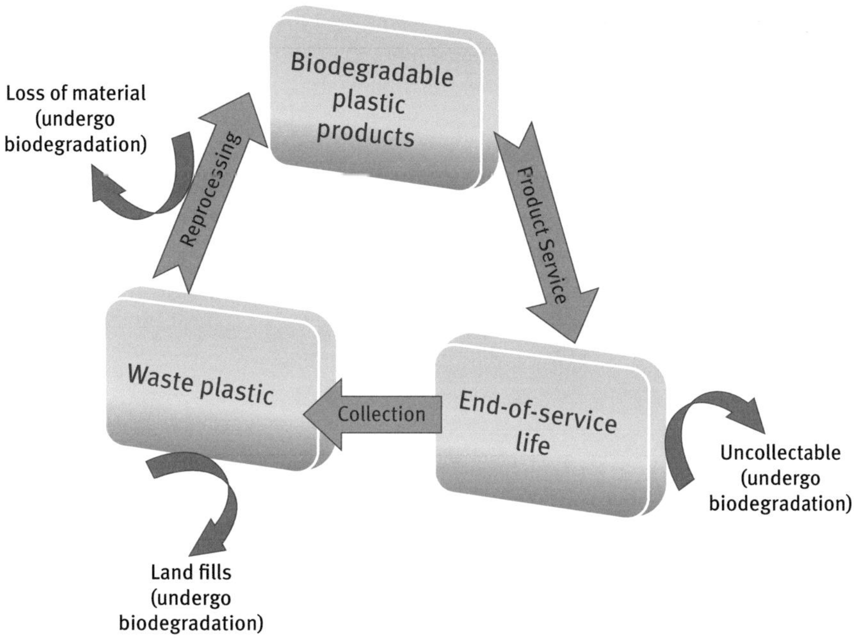

Figure 11.1: A schematic illustration indicating the recycling of biodegradable plastic materials.

11.2.1 Mechanical recycling of bioplastics

Mechanical recycling is one of the simplest and best energy-saving technique for reusing PLA wastes through multiple reprocessing under the same system in which the primary product is produced. Reprocessing is mostly done under high temperature and shear, which leads to the material degradation when it is subjected to multiple cycles. Therefore, the impact of thermomechanical degradation upon multiple reprocessing on the important physical, mechanical and thermal properties of biodegradable plastics needs to be addressed in proper manner. Pillin and coworkers studied the rheological and mechanical properties of PLA, wherein PLA is subjected to seven reprocessing cycles using molding machine. On the basis of their study, stable tensile modulus was observed with the thermomechanical cycles, whereas the strain at break, stress and rheological properties were diminished. This significant deterioration can be related with a large decrease in the molecular weight. The possibility of maintaining the properties through molar mass stabilization is also discussed using two chain stabilizers Tropolone and Quinone. Quinone was observed to be an efficient stabilizer to trap free radicals and maintain the molecular weight at the same processing conditions due to which mechanical properties were observed to be improved with reprocessing [18].

In another study, poly (lactic acid) was examined for multiple extrusion up to 10 times and the effect has been monitored using static tension test and Charpy impact strength (for mechanical properties), thermal stability (trough phase transition temperatures, melt flow rate and degradation temperature) and barrier properties (permeability of water vapor and oxygen). According to the investigation done by Zenkiewicz and coworkers, it was found that the melt flow rate, water vapor and oxygen transmission rates gradually increased with increasing the number of the extrusion cycles. Impact strength clearly decreased with slight reduction in tensile strength. However, glass transition temperature was not affected by increasing in the number of extrusion cycles, whereas decrement in cold crystallization temperature and slight diminishing in the melting point were observed. Based on their investigation, they concluded that PLA waste can be considered as suitable to be reused as an additive to a neat polymer [19].

The impact of multiple extrusion on the structure and properties of amorphous polylactide (PLA), and the change in morphology, segmental dynamics, thermal and mechanical performance related to thermomechanical degradation were also addressed by Badia and coworkers. The effect of thermomechanical degradation prompted through five consecutive injection cycles was studied. PLA remained in amorphous form throughout the reprocessing cycles, and the data of cold crystallization (from DSC measurement) showed an increment of enthalpy at each reprocessing step, which caused due to the thermomechanical chain scission. The influence of such chain scission on the glass rubber relaxation properties were studied by dynamic mechanical analysis that showed an increase in free volume affecting the segmental dynamics of PLA in the second reprocessing step [20].

Duigou and coworkers had investigated the recyclability of flax/PLLA (poly (L-lactide)) biocomposite (20% and 30% fibers by weight) for the advantage of biocompostability of natural fiber-reinforced biopolymers, also known as *biocomposites*. Their main objective was to provide an attractive alternative to glass fiber-reinforced petrochemical polymers. Mechanical properties were evaluated initially, and shown to be similar to those of glass/PP composites. Under repeated injection cycles, tensile properties were shown to be conserved till the third cycle. Molecular weight, thermal and rheological measurement was performed to study the effect on mechanical properties, matrix degradation and reduction fiber aspect ratio. With injection cycles, they observed lower molecular weight, lower glass transition temperature, a decrease in fiber length and separation of fiber bundles from the matrix. However, the property retention even after three cycles (100% recycling with no added virgin polymer) showed promising recyclability of the materials [21].

11.2.2 Chemical/biological (enzymatic) recycling

The main factors for depolymerizing reaction are temperature, catalyst concentration, reaction medium and pressure. Possible reactions mechanisms are backbiting, intramolecular transesterification and intermolecular transesterification. Considering PLA, the final product type depends on temperature and catalyst concentration and the possible outcomes of the reaction are lactic acid L-lactide, D-lactide, meso-lactide, cyclic oligomers and randomly degraded polymer chains. Chemical recycling of PLA is generally achieved by hydrolysis, alcoholizes and thermal depolymerization.

11.2.2.1 Hydrolysis

Highest yield of L-lactic acid (96%) can be achieved through hydrolysis of PLA at high temperature (270 °C) within 8 min in batch system [22]. The efficiency of recovering the monomeric unit is highly dependent on the processing condition. The studies of by Tsuji and coworkers revealed that irrespective of the physical state of PLA (solid or melt state), depolymerization occurs due to bulk erosion mechanism at a temperature range of 120 °C–190 °C. The depolymerization times to recovery more than 95% LA in the solid state (120 °C) and the melt state (180 °C) were observed to be 4,000 min and 130 min, respectively. The lower temperature range (120 °C to 190 °C) resulted in a 100-fold increase in depolymerization times with respect to the higher temperature range (250 °C to 350 °C) [23]. Complete degradation/depolymerization of PLA was observed in aqueous sodium hydroxide solution (NaOH) media. About 8 g of NaOH per 10 g of PLLA is required to degrade PLA completely at a temperature of 180 °C and 160 °C within 20 min and 60 min, respectively [24]. It is important to note that this is a relatively large quantity of sodium hydroxide, which may be an inhibiting factor for scale-up of the process.

The main drawback of this technique is its high-pressure reaction condition, which makes it energy intensive. The use of basic and acidic medium can facilitate the depolymerization. However, its reaction mode (batch process) still renders it inefficient compared to the continuous process.

11.2.2.2 Alcoholysis

Substituting water, which has weak nucleophile nature by alcohols, has a tendency to facilitate the reaction and helps the depolymerization to take place at lower temperature and pressure relative to hydrolysis Small carbon chain alcohols and macromolecules of diols can be used as depolymerization agents [25–27]. By using macromolecules having more than one active sites, rapid depolymerization can be achieved in a controlled reaction condition and ionic liquids as a catalyst can improve the depolymerization rate as well as the recovery of the final product [28].

11.2.2.3 Enzymatic depolymerization

Protease- and lipase-type enzymes have been tested to transform poly(lactic acid) in to its cyclic oligomers in organic solvents [29]. Takahshi and coworkers investigated the use of lipase RM (Lipozymes RM IM) and the lipase CA (Novozym 435) to catalyze the depolymerization reaction of PDLLA and PLLA, respectively, into cyclic oligomer. Reaction time, temperature, polymer concentration and solvent type are observed to be the controlling factors of the enzymatic degradation process [30]. The strategy of depolymerization of stereocomplex PLA into enantiomeric pure form of poly(L-lactic acid) and poly (D-lactic acid) was developed by Kawati and coworkers. The stereocomplex (sc-PLA) is formed by blending 50:50 of PLLA and PDLA and enantioselective degradation studies that occurred in amorphous region of PLLA/PDLA film and PDLA preferentially hydrolyzed by lipase-type (CLE) [31]. Enzymatic depolymerization takes place relatively at low temperature and stereoselective lactic acid can be easily achieved when compared to hydrolysis, alcoholysis or thermal depolymerization. Due to this, enzymatic depolymerization can be termed as *not energy intensive*. However, the reaction rate is very slow compared to the others techniques.

Effect of reaction system needs to be selected on the basis of purity and yield of monomeric unit at minimized temperature and time. Reaction rate is controlled by the diffusion of depolymerization agent (solvent/ catalyst/initiators) into the polymer chain and its activity on macromolecular level. Sufficient mixing environment is required for the reaction to provide maximum homogenization or good stoichiometry to complete the reaction in reasonably short time and temperature. Reaction medium such as alcohols, water and ionic liquids coupled with mechanical stirring, ultrasonication, microwave radiation or supercritical condition increases the rate of depolymerization. Temperature, pressure, initiators and PLA/catalyst ratio influence yield of the final product. Recently microwave reactors were observed to play a vital role in a number of chemical reactions. Microwave provides a concentrated energy to the reactant that helps in minimizing the reaction time and energy significantly compared to the conventional techniques [26].

11.2.2.4 Thermal depolymerization

PLA chains open up from the end by back-biting reaction and release lactides in a controlled unzip depolymerization above 250 °C. Compounds (alkoxides, organic acid and enolate salts) of Al, Ti, Zn and Zr have been evaluated to substitute stannous 2-ethylhexanoate (Sn(oct)2) [32, 33]. The purity of lactides was observed to be dependent on the catalyst concentration and temperature [34, 35]. US patent filed by Narayan and coworkers in 2013 describes the mechanism of thermal depolymerization of PLA, with the possibility of batch and continuous production of lactides [36]. The invention provided direct production method of lactide from PLA waste using thermal depolymerization. Generally, the method comprises of the following stages: preheating, depolymerization and lactide recovery. PLA feedstock is fed into a PLA preheater, where the temperature is maintained at 190 °C (above PLA melting temperature 173–178 °C.). After melting, PLA liquid was pumped into PLA depolymerization reactor. The temperature of the reactor was maintained at 200 °C and the reactor headspace was maintained under near vacuum condition (0.1 bar). Catalysts like Ti (ll) ethyl hexanoate (but not limited to it) are used to depolymerize PLA in to lactide. Lactide produced in the form of vapor recovered continuously from this stage and changed into liquid state in the condenser at 100 °C, which is above the melting point of lactide (96 °C). They proposed the process to be operated with either continuous feeding mode (PLA feed stock is fed to the system continuously via the preheater) and then PLA liquid is fed to the reactor from where the lactide recovered continuously or batch fed mode (with fed batch mode, the PLA liquid is fed to the reactor till 1/2–1/3 reactor volume, the feed is then stopped and the lactide is recovered continuously). They performed the experiment to prove their innovation. PLA feedstock was melted in a preheater at 170 °C and the liquid PLA intermediate transferred in to the reactor, then depolymerization occurred in the presence of Ti (ll) catalyst (0.05–0.1%,W/W) at a temperature range between 195–200 °C. They reported that the conversion efficiencies of PLA to lactide were greater than 91–95% (W/W) and the products were mainly L-lactide, with less than 10% of meso-lactide and cyclic oligomers [36].

11.3 Recycling of other biodegradable plastics

11.3.1 Polyhydroxybutyrate (PHB)

Recycling study of neat biopolymers other than PLA is scanty in report. Researchers have claimed to have mechanically recycled PHA with a slight reduction in molecular weight and mechanical properties [37]. Studies reported by Zaverl and coworkers showed that mechanical recycling of polyhydroxybutyrate-*co*-valerate did not cause significant change in mechanical properties up to five cycles [38, 39]. Chemical recycling was also reported in the case of PHA, where thermal degradation caused the formation of vinyl monomers while enzymatic transformation led to the formation of oligomers [13, 14].

11.3.2 Polybutylene ssuccinate (PBS)

PBS is a promising biodegradable material that can be recycled without much difficulty. The mechanical properties of PBS were found to be increased after reprocessing, which cannot be observed in the case of other plastics. Hence, PBS is a biodegradable material that can be recycled and thereby reduce the overall environmental load through waste management. Limited reports are available on recycling study of PBS. Ding and coworkers were capable of depolymerizing PBS via enzymatic hydrolytic degradation using lipase as catalyst. In the report, they successfully degraded high molecular weight PBS into small oligomers of molecular weight from few hundred to thousands. This has made tremendous influence in the future prospects of recycling of PBS [15]. Recently in the year of 2016, Georgousopoulou and coworkers reported improved thermomechanical stability and recyclability of PBS at 190 °C on the addition of stabilizer Irganox® 1010, Irgafos® 168 at an optimized concentration of 0.1 wt% [16]. However, a further increment in the concentration of stabilizer affects the thermomechanical stability of PBS adversely. Okajima and coworkers reported enzymatic depolymerization of PBS using lipase CA (Novozym® 435) in water-containing toluene at 50 °C [40]. In the study, PBS of molecular weight M_w of 22,000 g/mol was converted into low molecular weight cyclic oligomers mainly consisting of dimers, which can be repolymerized into higher molecular weight PBS with M_w of 52,000 g/mol.

11.3.3 Cellulose acetate

Cellulose acetate is a natural plastic that is prepared from purified natural cellulose. Although it cannot be melt processed in neat form unlike other thermoplastics, but by mixing with suitable polymers or plasticizers or other additives, it can be processed in the extruder. In the year 2010, a patent on recycling of cellulose acetate was disclosed, where the cellulose acetate waste was hydrolyzed by shearing and heating a combination of a hydrolyzing agent and cellulose acetate [7]. Distillation of hydrolyzing agent leads to the formation of a mixture of monosaccharides, polysaccharides, partially hydrolyzed cellulose and acetic acid.

11.3.4 Polycaprolactone (PCL)

Although mechanical recycling of PCL is not studied much, but the chemical recycling of PCL is studied by several researchers. Noted improvement in molecular weight was reported in specific to enzymatic recycling. Ebata and coworkers in the year 2000 reported effective depolymerization of PCL of M_n of 110,000 into corresponding cyclic dicaprolactone (DCL, 1,8 dioxacyclotetradecane-2,9-dione), with the yield of 97% using Candida antarctica lipase (lipase CA) in water-containing toluene at 40 °C [11]. Further, enzymatic transformation of PCL into PCL having an M_n of 110,000 into repolymerizable caprolactone (CL) oligomers in supercritical carbon dioxide ($scCO_2$) using Candida antarctica lipase (lipase CA) at 40 °C was reported by Matsumura and coworkers [8]. These CL oligomers can be easily repolymerized to high molecular weight PCL using the same enzyme.

11.4 Composting

Recent trends show that a large number of products known as *biodegradable*, and compostable are developed with the purpose of sustainable applications. Biodegradable materials that are commonly present in the market are PLA, PVA, PHB, PBAT [poly(butylene adipate-*co*-terepthalate)] and some other commonly known biodegradable plastics are cellulose. Most of these plastics are specifically targeted toward packaging applications so that they may help facilitate scrap collection for their composting. Collection of these scraps and commercial composting facilities are increasing day by day in different parts of the world with increasing use of biodegradable plastics. However, nowadays many researches have also targeted specific microbial strains that can degrade conventional plastics and have succeeded up to certain extent. This part of the chapter will provide a brief overview about compostable plastics and will define various basic terms such as composting characteristics and its significance, as well as will also highlight the key challenges and maximum opportunities of these plastics. The section will also provide solutions to the key questions and foster an intelligent answer to the basis of compostability of plastics. Petroleum-based products are the major raw materials used to manufacture most of the plastics. Replacing fossil-based plastics with renewable raw material-based biodegradable plastics will certainly reduce our dependency on fossil fuels [8, 9, 41]. Design and development of biodegradable or compostable plastics can provide enormous benefits to the environment. The use of biobased as well as compostable plastics will reduce our dependency on fossil fuels, which will in turn increase the demand for more sustainable products. Nevertheless, these kinds of plastics also create confusion among stakeholders due to inconsistent product labeling and definitive industry terms and how to use these products might be confusing for consumers and hence presents challenges to create large uncertainty for wide array of stakeholders. If biodegradable plastics are not sorted out properly, they can create problems during recycling and uneven decomposition rates during composting process and subsequently might create problems for industrial composting operations. A wide group of biodegradable plastics are developed for food packaging applications that have functionalities and processabilities comparable to conventional petrochemical-based plastics [42, 43]. The degree of biodegradation of such plastic materials largely depends on the chemical structure of the material rather than the resource of a material. In other words, all biobased plastics may not be biodegradable, and all fossil-based plastics may not be nonbiodegradable. Currently, biodegradable plastics represent about ~1% of total plastics produced annually, which will be ~300 million tons of annual production of plastics. However, with an increasing demand in sophisticated materials, its applications and products being researched, the market growth is about 20 to 100% per year. Recent market statistics compiled by European Biodegradable Plastics, production of biodegradable plastics will increase from ~1.7 million tons in 2014 to ~7.8 million tones by the year 2019. PHA production will be doubled by the year 2019 from 2014 [13].

11.4.1 Difference between bioplastics and compostable plastics

Plastic materials originating or produced from biosources such as corn, starch, sugar and so on, unlike conventional fossil-based sources, are known as biobased plastics. Biobased plastics are also known as *plant-derived products*. Hence, plastic that biodegrades within certain time frame is called *biodegradable plastic* [42]. According to the ASTM standard, compostable plastics undergo biodegradation via biological processes and in this case composting yields CO_2, H_2O and different inorganic compounds. Thus, compostable materials leave no mark of any visible and distinguishable or toxic residue [43] and convert 90% carbon content of the material into CO_2 [44].

11.4.2 Significance of composting of biodegradable plastics

Unlike recycling of bioplastics through processing, nature carries out recycling through composting. Composting leads to biodegradation of several organic wastes such as food, manure, leaves, grass, paper, wood and so on, turning them into numerous valuable organic fertilizers. Composting of biodegradable plastics is the biological process, which is generally conducted under controlled and aerobic conditions. This process breaks down organic matter into simpler substances under the action of various microorganisms, such as bacteria and fungi. The effectiveness of this process depends upon the environmental conditions and constituents of the composting system, that is, temperature oxygen, organic matter, moisture, the size and activity of microbial populations and material disturbance. In natural environment, composting occurs on a continuous basis and organic matter is metabolized by microorganisms, which is further consumed by invertebrates that in turn act as a nutrient to the soil and support plant growth. Composting is relatively very simple and can be carried out in different forms from indoor activity to outdoor or industrial composting process [45]. Composting of bioplastics can play a critical role in diverting waste from landfills, thereby reducing leachates and landfill spaces [10]. Various research interests in plastics that are biodegradable and might prove to be a boon for the packaging industries arise primarily from the use of renewable raw materials instead of fossil fuels and their excellent properties also paved way to end-of-life waste management by anaerobic digestion or composting, which reduce landfilling and promote sustainable development [26].

11.4.3 Principles of composting

Composting has the potential to transform biodegradable waste materials such as biodegradable plastics into fertile soil, enhancing the productivity of agriculture [44]. Biodegradation of natural compounds produce valuable compost as the prime product along with H_2O and CO_2 as by-products. In this process, the generated CO_2 does not affect the environment; CO_2 produced in this case does not contribute to an increase in greenhouse gases because the composting process is already a part of the biological carbon cycle.

11.4.4 Standards for compostability and biodegradability

Various international organizations like the ASTM and ISO have specified certain protocol for biodegradability tests of biodegradable polymer and some organic matters to confirm whether a product is biodegradable. In order to label a product biodegradable or compostable, the company/laboratory/organization must follow certain specified guidelines fixed according to different environments. Some of the international standards and legislations used for biodegradability/composting of biodegradable plastic materials are given in Table 11.1. Some of the companies/agencies that provide certification for compostable packages are as follows: Association Francaise de Normalisation (AFNOR) is a France-based certifying company. The Biodegradable Products Institute (BPI) is another not-for-profit association of key individuals and groups from government, industry and academia. They give certification for compostable packaging as well as products with a unique labeling logo for marketing and selling the product in the market.

Organizations like ASTM have developed certain tests that deal with the disposal by establishing guidelines, which suggest whether a certain bioplastic will biodegrade in a particular environment or not. A good research laboratory must

Table 11.1: Some of the widely used biodegradability/compostability test methods.

Standard	Title of the test method	Conditions	Objective biodegradation/ composting	Application
ASTM D 5210	Test method for determining the anaerobic biodegradation of plastic materials in the presence of municipal sewage sludge	Aerobic	Biodegradation	Packaging
ASTM D 5247	Test method for determining the aerobic biodegradability of degradable plastic by specific microorganisms	Aerobic	Biodegradation	Plastic
ASTM D 5271	Test method for determining the aerobic biodegradation of plastic materials an activated sludge wastewater treatment system	Aerobic	Biodegradation	Plastic
ASTM D 5338	Test method for determining aerobic biodegradation of plastic materials under controlled composting conditions, incorporating thermophilic temperatures	Aerobic	Composting	Plastic
ASTM D 5509	Practice for exposing plastics to a simulated compost environment.		Composting	Plastic
ASTM D 5511	Test method for determining anaerobic biodegradation of plastic materials under high solids anaerobic-digestion conditions	Anaerobic	Biodegradation	Plastic
ASTM D 5512	Practice for exposing plastics to a simulated compost environment using an externally heather reactor		Composting	Plastic
ASTM D 5526	Test method for determining anaerobic biodegradation of plastic materials under accelerated landfill conditions	Anaerobic	Biodegradation	Plastic
ASTM D 5988	Test method for determining aerobic biodegradation in soil of plastic materials or residual plastic materials after composting	Aerobic	Biodegradation in the soil	Plastic
ASTM D 6003	Test method for determining weight loss from plastic materials exposed to simulated municipal solid waste (MSW) aerobic compost environment		Composting	Plastic
ASTM D5209	Standard test method for determining the aerobic biodegradation of plastic materials in the presence of municipal sewage sludge	Aerobic	Biodegradation	Plastic

Table 11.1 (continued)

Standard	Title of the test method	Conditions	Objective biodegradation/ composting	Application
ASTM D5210	Standard test method for determining the anaerobic biodegradation of plastic materials in the presence of municipal sewage sludge	Anaerobic	Biodegradation	Plastic
ASTM E 1196	Test method for determining the anaerobic biodegradation potential of organic chemicals	Anaerobic		Chemical and organic products
ASTM D5929	Standard test method for determining biodegradability of materials exposed to municipal solid waste composting conditions by compost respirometry		Composting	
ASTM D 5975	Test method for determining the stability of compost by measuring oxygen consumption		Composting	
EN 13193	Packaging and the environment terminology			Packaging
EN 13427	Packaging requirements for the use of European standards in the field of packaging and packaging waste			Packaging
EN 13432	Packaging. requirements for packaging recoverable through composting and biodegradation. Test scheme and evaluation criteria for the final acceptance of packaging		Composting	Packaging
EN ISO 846	Plastics. Evaluation of the action of microorganisms			Plastic
ISO 10707	Water quality – evaluation in an aqueous medium of the ultimate aerobic biodegradability of organic compounds – method by analysis of biochemical oxygen demand (closed) bottle test	Aerobic aquatic media	Total biodegradability	Organic composition
AFNOR NF U 52-001	Biodegradable Mulching Film: test methods and criteria	Soil or aqueous media	Biodegradation	Mulching films

have to follow ASTM test methods to determine whether a product meets the ASTM standard for better research on global par. There are more than 20 active standards to carry out testing of biodegradability of plastics on the ASTM lists. Some of the primary ASTM test methods related to compostability of plastics are described below.

11.4.4.1 ASTM D5338

This is the standard protocol for testing how samples will biodegrade in a composting facility. The standard does not confirm the experiment as successful or unsuccessful, but provide it details about performing the experiment [44]. Figure 11.2 shows the process flowchart of material (PLA blend) conducted using a gas chromatography.

11.4.4.2 ASTM D6400

A set of three tests that includes ASTM D5338 must meet the standard of pass/fail criterion for judging the compostability of the plastics either in industrial or pilot plant-based composting facility. This criterion signifies that the product that has passed through this process can be termed/claimed to be compostable. Scope of this standard states that it covers plastics and various products made from plastics that are designed to be composted in both municipal and industrial composting facilities. Moreover, the properties of the composted product should not diminish the compost utility [13].

11.4.4.3 ASTM D7081

This standard is also the pass/fail protocol for attaining the understanding on compostability of plastic product in specific to the marine environment, such as seas and ocean. When the product passes this specification, it can be claimed to be biodegradable in marine waters. It is one of the most commonly used standard for testing biodegradability of the plastic products. Compostable plastics have tremendous food packaging applications accounting for more than 40% uses in conventional plastic production [46].

11.4.5 Measurement of biodegradability

Quantification of biodegradability is utmost essential to say if a product is biodegradable or not. Quantitative methods help in establishing disposal policies for

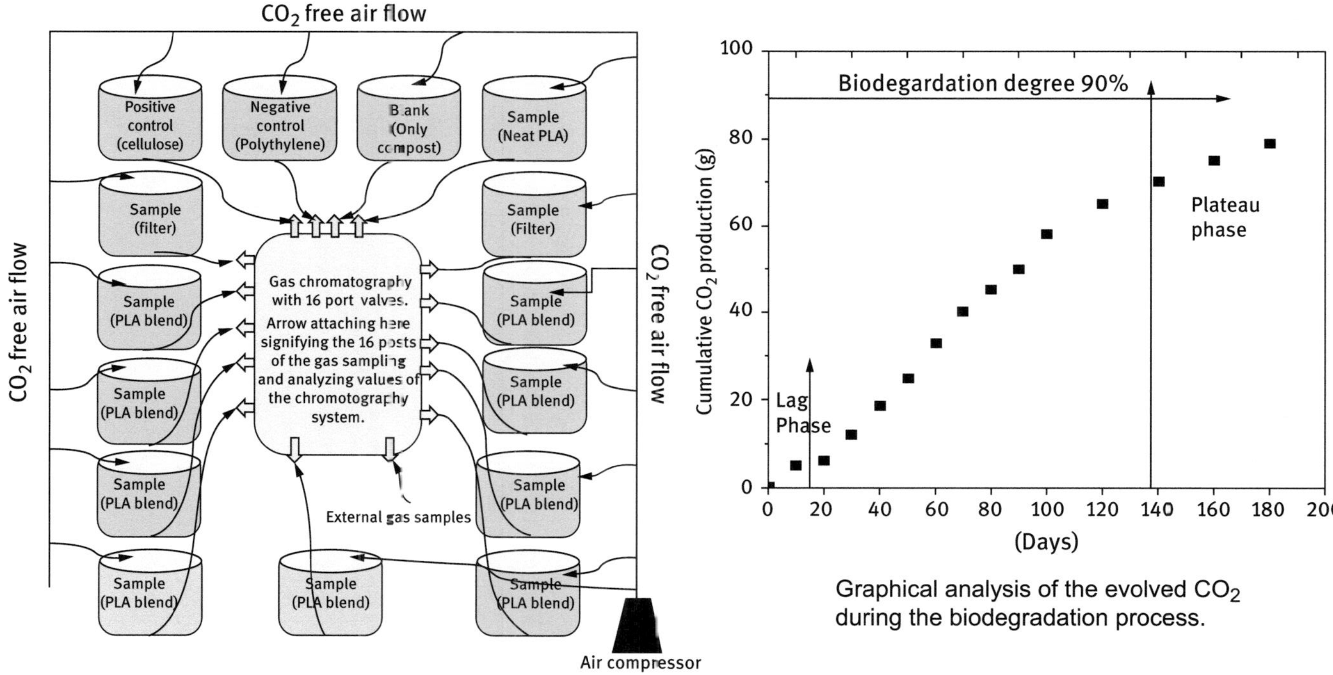

Figure 11.2: A flowchart of composting work of PLA blends in controlled aerobic environment based on ASTM D5338-15 (CoE-SusPol).

various packaging plastic materials. For addressing issues related to biodegradation, various test methods have been developed in recent times based on the product type and application. Biodegradation occurs in two different steps: aerobic biodegradation and anaerobic biodegradation. The mass balance equation for biodegradation that occurs in aerobic environment is given in eq. (11.1), which is characterized by total carbon conservation:

$$C_T = CO_2 + C_R + C_B \tag{11.1}$$

where C_T is total carbon content of the polymeric material, C_R is the residue left by the polymeric material or any other by-product formed during the degradation, CO_2 is the gaseous product that is measured out during the degradation process and C_B is the biomass produced by the microbes through growth and reproduction during the degradation process.

Similarly, the mass balance equation for biodegradation occuring in anaerobic environment is given by eq. (11.2), which is characterized by total carbon conservation:

$$C_T = CO_2 + CH_4 + C_R + C_B \tag{11.2}$$

Here, the gaseous products formed are mainly CO_2 and methane. In certain environments, both the aerobic and anaerobic biodegradation can take place simultaneously. Various other test methods have been established based on the above two equations, which quantify the biodegradability of polymer materials [43]. PLA, PBAT, PHB and cellulose are some of the most potentially rich and extensively studied biopolymers, which have been on the priority list of the scientific community for the commercialization in packaging sectors. Two important steps of biodegradation associated with polymers are hydrolysis of the polymers and its subsequent assimilation by microbes [8, 9, 41]. The use of plastic materials increases problems associated with waste disposal. The growing environmental impact of nonbiodegradable plastics has directed research to the development of plastics that degrade rapidly owing to biodegradation. PLA, a linear aliphatic polyester, is the one that has significant potential to replace the conventional petroleum-based plastics [9, 47, 48]. Microorganisms play a key role in the biodegradation of polymers in the real-environment conditions [41]. Various researches have been done based upon PLA as packaging material due to its biodegradability and processability [46]. A representative flow diagram of such nature to nature approach is shown in Figure 11.3 for the better understanding of the readers.

Researchers have investigated PLA films as well as PLA-based blends because blending offers wide scope of research for further improvement of PLA-based packaging systems [49–53]. Various researchers have investigated the biodegradation behavior of PLA and its blends under different environments. Weng and coworkers [54] investigated and found that P(3HB,4HB) and PLA have the fastest degradation rate in 20 and 40 cm soil depth, respectively. They also found that their blends have different degradation rates according to the different depths of soil where there were

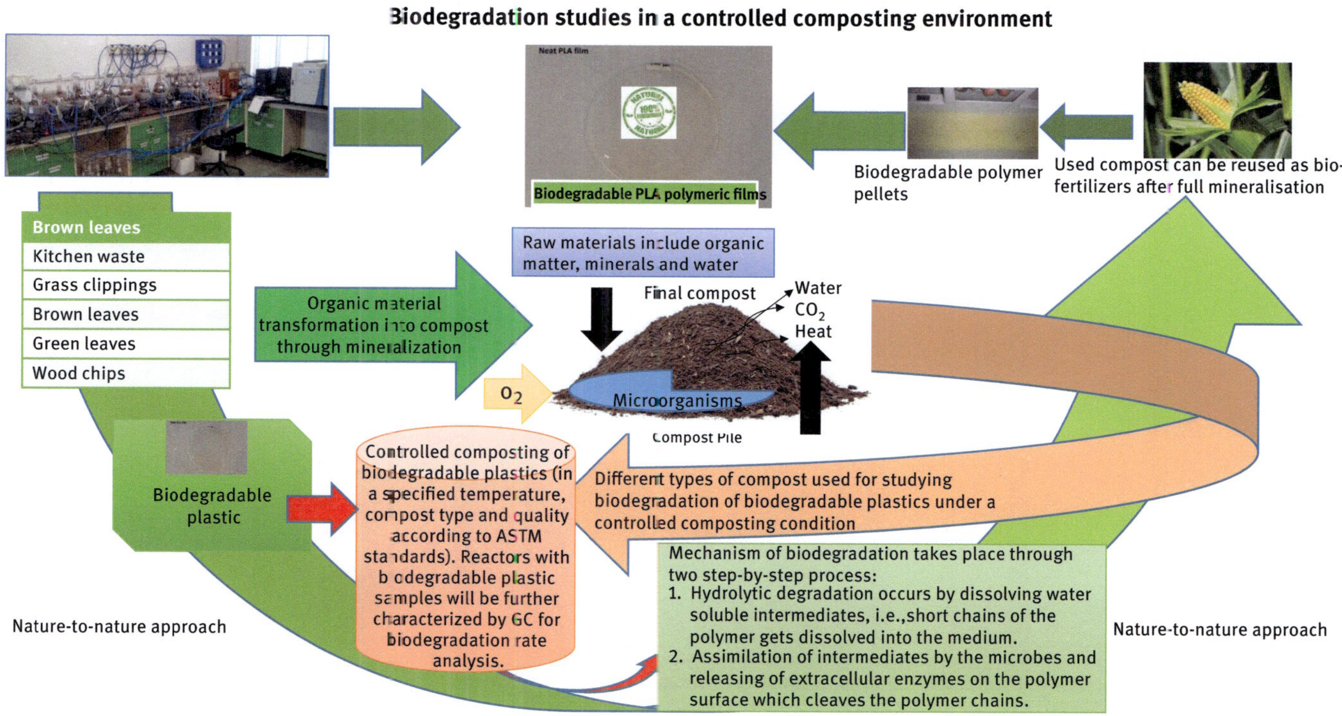

Figure 11.3: Composting “a nature-to-nature approach” (images are part of the research work carried out at CoE-SusPol, IIT Guwahati).

different humidities and microbial strains. Ghorpade and coworkers [55] demonstrated that composting of extruded PLA is an efficient process if added in less amount (mostly less than 30% by weight) to the precomposted waste under laboratory condition. Weng and coworkers [56] further investigated that the rate of biodegradation of PBAT as well as PLA in the PBAT/PLA blends is different from virgin materials, which further signifies that different composite blends would have different biodegradation rate based on the same biopolymer. However, investigations done by Bonhomme and coworkers [57] demonstrated that microbial growth observed on the polyethylene surface was attributed to the formation of protein and polysaccharides on the polymer surface, which initially leads to biofilm formation on the polymer surface, ultimately leading to biodegradation. Husarova and coworkers [49] isolated a specific PLA degrading bacteria and tentatively designated Thermopolyspora flexuosa and suggested that abiotic hydrolysis acts like a rate-limiting step in the biodegradation process of PLA, which further significantly lowers the rate of depolymerization due to lack of enzymatic activity of the concerned microbes. Various results showed that PLA is biodegradable both under composting condition as well as in real-environment conditions. Kale and coworkers [50] further found that PLA needs either more time or higher temperature for degradation.

Although there are various reports of biodegradation studies of PLA and PLA-based blends, biodegradation behavior differs according to different environments it is subjected to. Walczak and coworkers [51] investigated that respiratory activity is an indication that demonstrated different biodegradation rate of PLA under different environments due to the oxygen consumption by microbes under different environments. They found that oxygen consumption was higher in compost and the lowest in lake water.

A degradation study conducted by Orhan and coworkers, in 2004, used compost bags that were made of degradable polyethylene and nondegradable LDPE and HDPE [52]. The degree of biodegradation was measured in the soil that was mixed with 50% (w/w) manure community solid waste compost supplied by a municipality plant. Film samples were properly buried for 15 months under standard temperature in a 2 L capacity desiccator that contains soil, which has the capability of holding ~40% of its maximum water. The biodegradation of plastics was examined through gravimetric weight loss of samples, its mechanical property measurement, investigating its chemical changes in infrared spectrum, carbon dioxide production and overall microbial activity in the soil. In the experiment, high resistance to soil conditions was observed in the case of HDPE [51]. Moreover, such degradation might be affected from several environmental factors such as time, temperature, moisture and biological activity. Some environmental problems are also associated with composting since it releases CO_2 to the environment, which is quite less in quantity. In the composting study, biodegradable plastics were mixed with composting raw materials and during the composting of waste water sludge reduced NH_3 emission. They described biodegradable plastic might act as "reserve acid"; it is not acid itself but degrades and releases acid intermediates during the composting process. The plastics used are neutralized, which fixed the NH_3 generated decomposition of organic materials in the compost, which makes it possible in the reduction of NH_3 emission in composting process of the waste-water sludge [52]. Another study by Cadar and coworkers, in 2012, found that biodegradability of tested materials depends upon the type of processing of the polymeric material and monomeric content of the material [53]. Table 11.2 lists some of the biodegradable plastics and their uses in day-to-day life of our society.

Table 11.2: List of some of the biodegradable plastics and their uses.

Plastics	Uses	Composting Period
PGA	Controlled drug releases; Implantable parts or composites	N.A.
PLA	Packaging and coatings, mulch films, and compost bags	180 Days
PCL	Mulch and other agricultural films; slow release systems for drugs	60 Days
PHB	Products like bottles, bags, wrapping film and disposable napkins	N.A.
PHBV	Films and paper coatings; sustainable release systems	N.A.
PVOH	Packaging Application and used as material in detergents etc.	N.A.
PVAc	Packaging applications include boxboard manufacture, paper bags.	N.A.

Note: PGA: Polyglycolic acid; PLA: Polylactic acid; PCL: Polycaprolactone; PHB: Polyhydroxybutyrate; PHBV: Polyhydroxyvalerate; PVOH: Polyvinyl alcohol; PVAc: Polyvinyl acetate.

Overall composting of plastic is a nature-to-nature approach. Compost used for biodegrading biodegradable plastics can be reused as compost for agricultural purpose, without having presence of any toxic material in compost. Composting also reduces landfill hazards and landfill space can be utilized for other purposes. Biodegradable plastics will also reduce our dependence on fossil fuels. Composting is a well-defined end-life process cycle for biodegradable products, which has high-end applicability and a user-defined module for processabilty of the end products. Since biodegradable polymers are most suited for biological waste treatment through composting both in aerobic as well in anaerobic conditions, biodegradable polymers will play a bigger role in the packaging sector in the near future. Disposal of organic waste in the landfill conditions releases methane, which makes it unsuitable for landfill, and their disposal methods are also inconsistent, but to overcome all these issues composting proves to be a boon as composting of organic wastes will ensure nontoxic end products and will act as solid end-life cycle for biodegradable products. This chapter highlighted some key norms concerning labeling and testing of compostable plastics that have been developed and followed at national and international levels. This norm covers different product types. The media and testing facilities designed in the norms are mainly for industrial composting purposes. The basic biodegradability quantification criteria under composting conditions, such as the threshold percentages for biodegradation and disintegration, are presented in the norms. Based on those norms, various research activities were discussed and conducted both in controlled and uncontrolled environment for biodegradation of biodegradable plastics.

11.5 Challenges and future directions

Biodegradable plastics may be biobased – renewable or synthetic polymers that are produced either by using biomass through biological and chemical transformation or through synthetic routes with a tendency to degrade naturally in to carbon dioxide and water [58]. Due to this reason, biodegradable plastics are considered as a sustainable polymers as they have a potential to reduce fossil fuel consumption, minimize the waste disposal problem and emit less carbon dioxide during production, but the following major limitations restrict the mass production.

11.5.1 Feedstock

In the current scenario, food crops such as corn, soy and sugar cane are used as a feedstock for the production of biobased biodegradable plastics. Therefore, switching to biobased biodegradable plastics have the potential to affect the world's food supply. Biodegradable plastics derived from nonfood crops would indirectly affect the food production as well as its supply [59]. However, biobased biodegradable plastics derived from agricultural waste or food processing waste would have almost zero impact on the food supply [60].

11.5.2 Performance limitation

When compared to convectional polymers, biodegradable plastics have low thermal stability, mechanical and barrier properties, which reduce material applicability. The glass transition temperature and heat distortion temperature should be improved to expand its utilization [57].

11.5.2.1 Product cost

At present, the cost of bioplastics is higher than that of convectional polymers. If we consider PLA as an example, the cost is 3–5 times more than the cost of PET. It is obvious that other advantages of bioplastics are meaningless if the material is too expensive. Generally, bioplastics are superior to petroleum-based plastics in terms of energy efficiency, consumption of petroleum and carbon emissions, while low in cost and applicability span with an adverse impact on the food supply. Among all the synthetic biopolymers, poly(lactic acid) is the first in terms of production and applicability. In this particular research work, PLA is selected as a system to study the possible recycling alternatives since it accounts a large part of the postconsumer bioplastic waste, with promising future of expanding its applicability toward the substitution of conventional polymers.

References

[1] Sanchez, E M S. Ageing of PC/PBT blend: Mechanical properties and recycling possibility. Polym Test 2007, 26(3), 378–387.

[2] Zaverl, M., Seydibeyoğlu, M O., Misra, M., & Mohanty, A. Studies on recyclability of polyhydroxybutyrate-co-valerate bioplastic: multiple melt processing and performance evaluations. J Appl Polym Sci 2012, 125, E324–E331.

[3] Geyer, R., Jambeck, JR., Law, KL. Production, use, and fate of all plastics ever made. Sci Adv 2017 1;3(7):e1700782.

[4] Duigou, A L., Davies, P., & Baley, C. Life cycle analysis of a Flax / PLLA biocomposite. Matériaux & Techniques 2010, 98(2), 143.

[5] Badia, J D., & Ribes-Greus, A. Mechanical recycling of polylactide, upgrading trends and combination of valorization techniques. Eur Polym J 2016, 84, 22–39.

[6] Pillin, I., Montrelay, N., Bourmaud, A., & Grohens, Y. Effect of thermo-mechanical cycles on the physico-chemical properties of poly (lactic acid). Polym Degrad Stab 2008, 93(2), 321–328.

[7] Nishida, H., Fan, Y., Mori, T., Oyagi, N., Shirai, Y., & Endo, T. Feedstock recycling of flame-resisting poly (lactic acid)/aluminum hydroxide composite to L, L-lactide. Ind Eng Chem Res 2005, 44(5), 1433–1437.

[8] Ying, L., Rongqing, W., Jun, W., & Xiaoning, L. Thermal degradation of poly (L-lactide) with high molecular weight into L-lactide. Prog in Chem 2008, 20(10), 1588–1594.

[9] Fan, Y., Nishida, H., Shirai, Y., & Endo, T. Control of racemization for feedstock recycling of PLLA. Green Chem 2003, 5(5), 575–579.

[10] Ebata, H., Toshima, K., & Matsumura, S. Lipase-catalyzed transformation of poly (ε-caprolactone) into cyclic dicaprolactone. Biomacro 2000, 1(4), 511–514.

[11] Zaverl, M., Valerio, O., Misra, M., & Mohanty, A. Study of the effect of processing conditions on the co-injection of PBS/PBAT and PTT/PBT blends for parts with increased bio-content. J Appl Polym Sci 2015, 132, 2.

[12] Ding, M., Zhang, M., Yang, J., & Qiu, JH. Study on the enzymatic degradation of PBS and its alcohol acid modified copolymer. Biodegradation 2012, 23(1), 127–132.

[13] Georgousopoulou, I N., Vouyiouka, S., Dole, P., & Papaspyrides, CD. Thermo-mechanical degradation and stabilization of poly (butylene succinate). Polym Degrad Stab 2016, 128, 182–192.

[14] Okajima, S., Kondo, R., Toshima, K., & Matsumura, S. Lipase-catalyzed transformation of poly (butylene adipate) and poly (butylene succinate) into repolymerizable cyclic oligomers. Biomacromolecules 2003, 4(6), 1514–1519.

[15] Matsumura, S., Ebata, H., Kondo, R., & Toshima, K. Organic solvent-free enzymatic transformation of poly (ε-caprolactone) into repolymerizable oligomers in supercritical carbon dioxide. Macromol Rapid Commun 2001, 22(16), 1325–1329.

[16] Piemonte, V. Bioplastic wastes: the best final disposition for energy saving. J Polym Environ 2011, 19(4), 988–994.

[17] Pillin, I., Montrely, N., Bourmaud, A., & Grohens, Y. Effect of thermo-mechanical cycles on the Physico-chemical properties of poly (lactic acid). Polym Degrad Stab 2008, 932, 321.

[18] Żenkiewicz, M., Richert, J., Rytlewski, P., Moraczewski, K., Stepczyńska, M., & Karasiewicz, T. Characterisation of multi-extruded poly (lactic acid). Polym Test 2009, 28(4), 412–418.

[19] Badia, J D., Strömberg, E., Karlsson, S., & Ribes-Greus, A. Material valorisation of amorphous polylactide. Influence of thermo-mechanical degradation on the morphology, segmental dynamics, thermal and mechanical performance. Polym Degrad Stab 2012, 97(4), 670–678.

[20] Le Duigou, A., Pillin, I., Bourmaud, A., Davies, P., & Baley, C. Effect of recycling on mechanical behaviour of biocompostable flax/poly (l-lactide) composites. Composites Part A 2008, 39(9), 1471–1478.

[21] Saeki, T., Tsukegi, T., Tsuji, H., Daimon, H., & Fujie, K. Depolymerization of poly (L-lactic acid) under hydrothermal conditions. Kobunshi Ronbunshu 2004, 61(11), 561–566.

[22] Tsuji, H., Saeki, T., Tsukegi, T., Daimon, H., & Fujie, K. Comparative study on hydrolytic degradation and monomer recovery of poly (L-lactic acid) in the solid and in the melt. Polym. Degrad Stab 2008, 93(10), 1956–1963.

[23] Yagihashi, M., & Funazukuri, T. Recovery of L-lactic acid from poly (L-lactic acid) under hydrothermal conditions of dilute aqueous sodium hydroxide solution. Ind Engg Chem Res 2010, 49(3), 1247–1251.

[24] Plichta, A., Lisowska, P., Kundys, A., Zychewicz, A., Dębowski, M., & Florjańczyk, Z. Chemical recycling of poly (lactic acid) via controlled degradation with protic (macro) molecules. Polym Degrad Stab 2014, 108, 288–296.

[25] Hirao, K., Nakatsuchi, Y., & Ohara, H. Alcoholysis of poly (L-lactic acid) under microwave irradiation. Polym Degrad Stab 2010, 95(6), 925–928.

[26] Cadar, O., Paul, M., Roman, C., Miclean, M., & Majdik, C. Biodegradation behaviour of poly (lactic acid) and (lactic acid-ethylene glycol-malonic or succinic acid) copolymers under controlled composting conditions in a laboratory test system. Polym Degrad Stab 2012, 97, 354–357.

[27] Valapa, R., Hussain, S., Iyer, P K., Pugazhenthi, G., & Katiyar, V. Non-isothermal crystallization kinetics of sucrose palmitate reinforced poly (lactic acid) bionanocomposites. Polym Bull 2016, 73, 21–38.

[28] Valapa, R., Hussain, S., Iyer, PK., Pugazhenthi, G., & Katiyar, V. Non-isothermal crystallization kinetics of sucrose palmitate reinforced poly (lactic acid) bionanocomposites. J Appl Polym Sci 2015, 3, 132–139.

[29] Yutaka, T., & Buenaventurada, P C. Biodegradability and biodegradation of poly(lactide). Appl Micro Biotech 2006, 72, 244.

[30] Kaeb, H., European Bio-plastics: Introduction, www.european-bioplastics, accessed on November, 2016.

[31] Grala, A., Ejfler, J., Jerzykiewicz, LB., & Sobota, P. Chemoselective alcoholysis of lactide mediated by a magnesium catalyst: An efficient route to alkyl lactyllactate. Dalton Transactions 2011, 40(16), 4042–4044.
[32] Song, X., Wang, H., Yang, X., Liu, F., Yu, S., & Liu, S. Hydrolysis of poly (lactic acid) into calcium lactate using ionic liquid [Bmim][OAc] for chemical recycling. Poly Degrad Stab 2014, 1(10), 65.
[33] Watanabe, M., Kawai, F., Tsuboi, S., Nakatsu, S., & Ohara, H. Study on enzymatic hydrolysis of polylactic acid by endogenous depolymerization model. Macromol Theory Simul 2007, 16 (6), 619–626.
[34] Takahashi, Y., Okajima, S., Toshima, K., & Matsumura, S. Lipase-catalyzed transformation of poly (lactic acid) into cyclic oligomers. Macromol Biosci 2004, 4(3), 346–353.
[35] Kawai, F., Nakadai, K., Nishioka, E., Nakajima, H., Ohara, H., Masaki, K., & Iefuji, H. Different enantioselectivity of two types of poly (lactic acid) depolymerases toward poly (L-lactic acid) and poly (D-lactic acid). Polym Degrad Stab 2011, 96(7), 1342–1348.
[36] Noda, M., & Okuyama, H. Thermal catalytic depolymerization of poly (L-lactic acid) oligomer into L-lactide: Effects of Al, Ti, Zn and Zr compounds as catalysts. Chemi Pharma Bulletin 1999, 47(4), 467–471.
[37] Nishida, H., Mori, T., Hoshihara, S., Fan, Y., Shirai, Y., & Endo, T. Effect of tin on poly (L-lactic acid) pyrolysis. Polym Degrad Stab 2003, 81(3), 515–523.
[38] Fan, Y., Nishida, H., Hoshihara, S., Shirai, Y., Tokiwa, Y., & Endo, T. Pyrolysis kinetics of poly (L-lactide) with carboxyl and calcium salt end structures. Polym Degrad Stab 2003, 79(3), 547–562.
[39] Fan, Y., Nishida, H., Shirai, Y., Tokiwa, Y., & Endo, T. Thermal degradation behaviour of poly (lactic acid) stereocomplex. Polym Degrad Stab 2004, 86(2), 197–208.
[40] Narayan, R., Wu, W M., & Criddle, C S. inventors; Criddle C S, assignee; US20130023674 A1, 2013.
[41] Ariffin, H., Nishida, H., Hassan, MA., & Shirai, Y. Chemical recycling of polyhydroxyalkanoates as a method towards sustainable development. Biotechnol J 2010, 5 (5), 484–492.
[42] Kaihara, S., Osanai, Y., Nishikawa, K., Toshima, K., Doi, Y., & Matsumura, S. Enzymatic transformation of bacterial polyhydroxyalkanoates into repolymerizable oligomers directed towards chemical recycling. Macromol Biosci 2005, 5(7), 644–652.
[43] Ding, M., Zhang, M., Yang, J., & Qiu, J H. Study on the enzymatic degradation of PBS and its alcohol acid modified copolymer. Biodegradation 2012, 23(1), 127.
[44] Okajima, S., Kondo, R., Toshima, K., & Matsumura, S. Lipase-catalyzed transformation of poly (butylene adipate) and poly (butylene succinate) into repolymerizable cyclic oligomers. Biomacromolecules 2003, 4(6), 1514–1519.
[45] Combs, MT., & Woodson, BR., inventors; Celanese Acetate Llc, assignee. Process for recycling cellulose acetate ester waste. United States patent application US 12/686,619, 2010 Jul 15.
[46] Ravibabu, V., Hussai, S., Iyer, P. K., Pugazhenthi, G., & Katiyar, V. Non-isothermal crystallization kinetics of sucrose palmitate reinforced poly(lactic acid) bionanocomposites. Polymer Bulletin 2016, 73, 21–38.
[47] Yutaka, T., & Buenaventurada, P. C. Biodegradability and biodegradation of poly(lactide). App Micro Biotechnol 2006, 72, 244–251.
[48] What are bioplastics. (e.g. www.europeanbioplastics.org). accessed 19/10/2016 at 4:15 pm
[49] Long, Y., Katherine, D., & Lin, L. Polymer blends and composites from renewable resources. Prog Polym Sci 2006, 31, 576–602.

[50] Ohkoshi, I., Abe, H., & Doi, Y. Miscibility and solid-state structures for blends of poly((S)-lactide)with atactic poly((R, S)-3-hydroxybutyrate). Polymer 2000, 41, 5985–5992.
[51] Koyama, N., & Doi, Y. Miscibility of binary blends of poly[(R)-3- hydroxybutyric acid] and poly [(S)-lactic acid]. Polymer 1997, 38, 1589–1593.
[52] Zhang, L., Xiong, C., & Deng, X. Miscibility, crystallization and morphology of poly(b-hydroxybutyrate)/poly(D, L-lactide). Polymer 1996, 37, 235–241.
[53] Weng, Y X., Wang, L., Zhang, M., Wang, X L., & Wang, Y Z. Biodegradation behavior of P (3HB,4HB)/PLA blends in real soil environments. Polym Test 2013, 32, 60–70.
[54] Dhar, P., Bhardwaj, U., Kumar, A., & Katiyar, V. Poly (3-hydroxybutyrate)/cellulose nanocrystal films for food packaging applications: barrier and migration studies. Poly Engg and Sci 2015, 55(10), 2388–2395.
[55] Mistriotisa, A., Briassoulisa, D., Giannoulisa, A., & Salvatore, D A. Design of biodegradable bio-based equilibrium modified atmosphere packaging (EMAP) for fresh fruits and vegetables by using micro-perforated poly-lactic acid (PLA) films. Post Bio and Tech 2016, 111, 380–389.
[56] Wang, L F., Rhim, J W., & Seok-In, Hong. Preparation of poly(lactide)/poly(butylene adipate-co -terephthalate) blend films using a solvent casting method and their food packaging application. Food Sci & Technol 2016, 68, 454–461.
[57] Richards, E., Rizvi, R., Chow, A., & Naguib, H. Biodegradable composite foams of PLA and PHBV using subcritical CO_2. J of Poly & Environ 2008, 16, 258–266.
[58] Weng, Y X., Jin, Y J., Meng, Q Y., Wang, L., Zhang, M., & Wang, Y Z. Biodegradation behavior of poly(butylene adipate-coterephthalate) (PBAT), poly(lactic acid) (PLA), and their blend under soil conditions. Polym Tes 2013, 32, 918–926.
[59] Bonhommea, S., Cuerb, A., Delortb, A M., Lemairea, J., Sancelmeb, M., & Scottc, G. Environmental biodegradation of polyethylene. Polym Degrad Stab 2003, 81, 441–452.
[60] Kale, G., Auras, R., Singh, S P., & Narayan, R. Biodegradability of polylactide bottles in real and simulated composting condition. Polym Test 2007, 26(8), 1049–1061.

Gourhari Chakraborty, Purabi Bhagabati and Vimal Katiyar

12 Authors' view point toward developments in biodegradable polymers to improve their versatility in food packaging

12.1 Introduction

Plastics are commonly used as food packaging materials to maintain the food quality, to protect it from external factors such as oxygen, moisture, chemical, radiation and microbial attack. Polymers protect food from physical and external contamination so as to retain their texture, taste, flavor during storage, transportation and use. Petroleum-based plastics like polyethylene, polypropylene, polyethylene terephthalate (PET) and so on possess variation in barrier, mechanical and thermal properties and are being produced commercially since the mid of twentieth century [1]. But these conventional plastic materials pose a serious threat to environment due to their nondegradable nature; however, the hazards can be minimized by effective waste-management approaches. Another major concern is depleting petroleum feedstocks, which is the raw material for the production of these plastics, leading us to explore the biodegradable materials that could serve as precursors in producing various polymers. These biodegradable materials should be derived from abundant renewable resources, which may be fossil or biobased, but the structural and physical properties of biodegradable materials need to be modified so as to make them useful as sustainable products for potential replacement of petroleum-based plastics. The advantages of biodegradable plastics over conventional plastics include reduced toxic emissions, eliminating the use of depleting petroleum resource, environment-friendly processing, low energy consumption during production and an overall reduced carbon footprint. However, biodegradable polymers have not been extensively utilized in commercial food packaging applications due to their shortcomings in some properties and processability, as most of the biopolymers come under the category of plastics. Despite several limitations of biopolymers, various chemical and physical modifications can enhance their properties, making them acceptable for processing and therefore for food packaging applications. Structural properties of biodegradable materials can be enhanced by fabricating their nanocomposites, leading to their increased acceptability [2]. In the case of edible capsules, the use of nanofillers can lead to controlled drug release in selective organs such as heart, brain and so on. This chapter focuses on various limitations and potential of biodegradable polymers and possible modifications to enhance their properties as potential replacement of conventional packaging materials.

https://doi.org/10.1515/9783110648034-012

12.2 Limitations of biodegradable polymers as food packaging material

Plastic materials used in packaging applications are synthesized from different polymers having different chemical nature and physical structure and are selected for intended use based on their properties and processability. Biodegradable polymers such as starch, cellulose derivatives, polylactic acid (PLA), polycaprolactone (PCL) and polybutylene succinate are the most viable materials for different packaging applications. Properties and processing strategy of biodegradable polymers are two crucial parameters to be looked upon to enhance their versatility. The following section describes about the material properties of biodegradable polymers for packaging applications as compared to conventional petroleum-based plastics along with the modifications required to overcome the limitations.

12.2.1 Material properties

Properties of biodegradable polymers need to be assessed before using them as primary packaging materials. This includes mechanical properties, possible microbial growth, degradation rates under diverse conditions and migration of harmful substances into food products. To maintain the quality of packaged materials, it is important to maintain barrier toward oxygen, water vapor and aroma. Thermoplastic nature, developing thermal stability and heat deflection temperature of polymers are associated with melt processing and high-temperature applications.

12.2.2 Mechanical properties

The biodegradable polymers have inferior mechanical properties such as tensile strength, impact strength, Young's modulus, elongation at break, compression strength and tear resistance, which limit their use in packaging applications. The mechanical properties can be enhanced by fabrication of biobased nanocomposites that are made by reinforcing various biobased nanomaterials such as cellulose nanocrystals, sucrose palmitate, bamboo fiber and so on and inorganic materials such as calcium sulfate, silicates, clay, graphene, carbon nanotube (CNT) and so on. Valapa et al. in 2013 reported the improvement in tensile strength of PLA by up to 25% by reinforcing it with graphene [3]. Dhar et al. in 2015 reported improvement in strength of PLA by 74% by using cellulose nanocrystal as biofiller [4] Kathuria et al. in 2013 have reported the improvement in elongation at break and toughness of PLA by 5 and 3 times, respectively, as compared to neat PLA [5]. Different plasticizers like polyethylene glycol and soya bean oil have been reported for their use in improving elongation properties of biodegradable polymers.

12.2.3 Barrier properties

It is essential for food packaging materials to provide a barrier to oxygen, water vapor, aroma and different gases so that desired quality of food is maintained. Biodegradable polymers like PLA and PCL restrict the permeation of water vapor, but they are known to have high-oxygen transmission rate (OTR). PET is considered as the most commonly used commercial packaging material having extremely good oxygen barrier properties (O_2 permeability = 6–8 mmol m^{-1} s^{-1} GPa^{-1}). OTR of biodegradable polymers can be reduced by reinforcing them with CNT, starch, silicates and cellulose nanocrystals. Valapa et al. in the year 2015 reported that oxygen permeability of PLA was reduced by 69.5% on reinforcing it with sucrose palmitate [6]. Tripathi et al. in 2016 have utilized modified gum arabic as filler, which substantially reduced the oxygen permeability of PLA by 90% in the case of 5% loading [7]. Clay is also widely used as a modifier to improve the oxygen barrier properties of biodegradable polymers. Water vapor transmission rate is an essential factor in determining the quality of polymer to use for packaging application. Clay, a hydrophilic filler, is known to improve oxygen barrier property but cannot restrict water vapor transmission. Thus, it is important to select reinforcing material so as to meet the desired application. Tripathi et al. in 2016 showed that gum can reduce the water vapor permeability of PLA by 27% [7]. Using modified clay for reinforcement of polycaprolactum can reduce the water vapor transmission [2]. Aroma of the food, especially for flavored food items, should remain intact for several months. Layered reinforcing filler can enhance the barrier properties of the biocomposites toward aroma and different gases.

12.2.4 Thermal stability

Biodegradable polymers are known to have low thermal stability when compared to commercially available polymers such as low-density polyethylene, high-density polyethylene and so on. Poor thermal stability of biodegradable polymers leads to an increase in degradation and reduction in molecular weight at elevated temperature during processing, especially in the case of extrusion. Different carbon-based fillers like CNT, graphene, graphite and so on can increase the onset degradation temperature of biodegradable polymers. Murariu et al. in 2010 improved the onset degradation of PLA by 8 °C by incorporating 12 wt% expanded graphite [8]. Valapa and coworkers in 2013 reinforced exfoliated graphene into PLA, leading to the enhanced thermal stability of PLA [3]. Thermal stability can also be improved by using crosslinking agents in the case of biodegradable polymers.

12.2.5 Temperature resistance

Temperature resistance is another important factor that needs to be addressed so as to improve the application of biodegradable polymers in packaging. For instance, thermal stability of PLA only up to 100 °C is limiting its use in several packaging applications, especially those requiring high-temperature resistance. PLA is gaining wide acceptance in food packaging, which is an environmental friendly alternative to petroleum-based plastic. Pure PLA begins to soften at 60 °C, making it unsuitable for heat-intensive processes. Recently, the researchers at the Fraunhofer Institute for Applied Polymer Research, IAP in Potsdam, have developed heat-resistant PLA that can sustain up to 120 °C using PLA stereocomplex with L-lactides and D-lactides as main components [9]. Oyama and Abe in 2015 have made a breakthrough invention by making stereocomplexation and reactive blending of poly(L-lactic acid) (PLLA) and poly-D-lactic acid together with poly(ethylene-ran-methylacrylate–ran-glycidyl methacrylate), EMA-GMA [10].

12.2.6 Processing conditions

Casting polymers into films is possible using various methods depending upon end-use application. Different processing techniques are used to cast polymer films into useful products. Melt processing technique is the mostly utilized technique among all others for intended large-scale production. Processing biodegradable polymer films during extrusion encounters various degradation and crumpling problems. The properties of films are significantly affected by processing polymer melts under high shear and sudden cooling of the cast film, which may lead to depletion in molecular weight and crystallinity, thus leading to low commercial production rate of the biodegradable plastics. The applicability of biodegradable polymers can be enhanced by using bionanocomposites; however, the processability of nanocomposites is quite difficult due to the filler agglomeration and nonuniform dispersion. The presence of moisture can lead to severe deterioration in mechanical and thermal properties of PLA, because of thermally induced hydrolysis process taking place during processing. However, Dhar et al. have recently reported the development of thermally stable PLA-grafted cellulose nanocrystal (PLA-g-CNC) nanocomposite films by single-step reactive extrusion process that can be recycled multiple times without deterioration in its properties [11]. Hence, technical and scientific know-how can be helpful in overcoming the undesirable constraints in processing of certain biodegradable plastics.

12.3 Future horizons of biodegradable polymers as packaging materials

Biodegradable polymers suffer several limitations that can be overcome by using composites. In order to enhance the use of the biodegradable polymers in packaging applications, various materials can be reinforced in biodegradable polymers to manufacture biocomposites. These reinforcing materials can be particles or fibrous. Incorporation of reinforcing materials significantly improves barrier, mechanical and thermal properties, which ultimately improve the processability. Such advancements in biopolymers opens up new horizons for their applications in food packaging with smart properties such as sensing or antimicrobial function, which is termed as *active packaging*. Hence, "active packaging" is coming up as one of the major areas of significant research in food packaging. Along with bionanocomposites, blending and foaming of biodegradable polymers also lead to versatile food packaging materials. A new concept of "smart packaging" has been developed by the research group of the company "Sintef," where Larsson et al. reported the use of bionanocomposites of PLA and bio-PET, which not only improves the self-life of food products by preserving the food products from microorganisms, oxygen and moisture of air, but also has sensors indicating the deterioration of food [12]. According to Larsen et al., the biopolymer-based smart food packaging material contains some nanoparticles in the form of nanocapsules containing signal substances. At high temperature or at a pH value that is anomalous to the food substance, these nanocapsule shells decompose and release the signal substances. The sensors being highly sensitive undergo color change as the substances get released. In addition, composite laminates of a cellulose-based film sandwiched between two biodegradable polymer layers are also developed for improving oxygen barrier in food packaging applications.

12.3.1 Composites of biodegradable polymers

Biodegradable composites contain organic or inorganic substances depending upon the targeted property and field of application. Dimension of the reinforcing substance governs the compatibility with matrix polymer. Fabrication and utilization of the bionanocomposites contains at least one component having one dimension in nanometer range. Permeability is one of the most important parameters in the case of packaging applications. The barrier properties are enhanced significantly by incorporation of nanoclays. PLA, polyhydroxylbutyrate and starch-based nanocomposites are attractive packaging materials. Oxygen and water vapor barrier properties are reported to have been improved by 15–48% and 40–50% for PLA-clay nanocomposites compared to PLA. Barrier properties are also enhanced by reinforcing polysaccharides. Several other properties are also influenced and upgraded by fillers such as graphene, CNT and metal oxides.

12.3.2 Polysaccharide-reinforced composites

Polysaccharides like cellulose nanowhiskers and cellulose nanocrystal improve the degradation temperature, mechanical properties and crystallization nature. Onset degradation temperature was reported to increase with CNC content. For PLA, it increased to 292.2 °C, and for 10 wt.% composite to 305.4 °C. Young's modulus and elongation at break are reported to increase with CNC loading [11]. Microfibrilled cellulose/PLA composites fabricated using three-roll mill are reported to have higher elastic modulus and tensile strength of the composites with MFC loading [13]. CNCs are the promising nanomaterials that have the ability to form strong hydrogen-bonded network like structure within the polymer matrix, thereby transferring its stress to the nanocomposites efficiently. Compared to other nanofillers, high aspect ratio, high hydroxyl functionality and nontoxic effects aid in fabrication of nanocomposite films for food packaging applications.

12.3.3 Clay-reinforced composites

Clays are layered silicates consisting of tetrahedral silicon and octahedral metallic (Mg/Al) sheets fused together. Charge of layers depends on cation-exchange capacity and interlayer spacing rely on the type of the clay. Unmodified clay and organically modified clay can enhance barrier, flame-retardant properties and antimicrobial nature of the material. Incorporation of montmorillonite (MMT) clay can increase the decomposition temperature from 500 °C to 600 °C. Chitosan–MMT composites have been reported for better microbial activity than pure chitosan against *S. aureus* and *E. coli* [14]. PLLA and organically modified MMT (Closite® 20A and Closite® 30B) composites have good antimicrobial properties.

12.3.4 Carbon filler-reinforced composites

Carbon fillers can be of several types depending on the structure and properties of the allotropes. CNT, graphene, graphene oxide (GO), expanded graphite and so on having high mechanical strength in the order of TPa increases the strength of the composites. At 12% expanded graphite addition thermal stability of PLA is increased by 10 °C [15]. Tensile modulus is reported to increase by 172.8% and 204.9%, respectively, for composite made of poly(D, L-lactic-*co*-glycolic acid) (PLGA)/GO (1 wt%) and PLGA/GO (2 wt%) and glass transition temperature of 66.7 °C for 2 wt%

composite than PLGA and PLGA/GO (1 wt %) composites, for which the values were 32 °C and 33 °C [16]. Initial degradation temperature was reported to increase up to 280 °C from 200 °C in multiwall CNT reinforcement in PLA [17].

12.3.5 Metal- and metal oxide-reinforced composites

Metal-based composites are reported to have antibacterial activity efficiency of 98.5% and 99.2% against *S. aureus* and *E. coli* measured by viable cell count method in the case of Ag/PLA nanocomposites [18]. In the case of ZnO/PLA composite, antimicrobial activity enhanced and reduced 99% population of *S. aureus* and *E. Coli* [19].

12.3.6 Coated paper for active packaging

Paper has been widely used in various packaging applications including food packaging, which is biodegradable imposing no threat to the environment. However, it has high gas permeability due to its porous cellulose structure made of microfibrils. Paper easily absorbs water from environment as well as food due to the hydrophilic nature of cellulose, leading to deterioration of its physical and mechanical characteristics. However, it is common to coat such paper-based food-packaging materials with nonbiodegradable petroleum-based plastics like PE leading to loss of its biodegradability and recyclability. Nevertheless, biobased and biodegradable hydrophobic plastics materials have tremendous potential to replace the currently used synthetic paper coating plastics of paperboard [20, 21]. This biopolymer coating is capable of reducing the unwanted oxygen, moisture and oil transfer to the food products while retaining biodegradability of the material. Furthermore, incorporating various antimicrobial agents in the coating leads to the formation of active paper-packaging materials. The selection of active components is often limited by their incompatibility with the packaging material and therefore it is essential to select proper coating for the polymer matrix, active agents and plasticizers.

From the authors' view point, it is very important to understand the importance of biodegradable polymers in packaging applications. Biodegradable polymers have significant potential to be used as food packaging materials due to their versatile properties. Biodegradable polymers will not only replace the prevailing nonbiodegradable plastics but will also reduce the environmental load, leading to a better lifestyle and reduced health problems of our society. The recent advancements in food packaging applications include active food packaging, multilayer biodegradable plastic food packaging, edible packaging and edible coating, biodegradable foams in food packaging application and so on. Furthermore, bionanocomposites have the immense potential of replacing conventional plastic materials due to its versatile properties and biodegradability, eventually leading to overall benefit of both people and environment. Currently, tremendous research input has been focused to the development of biodegradable polymers for direct implementation in the application field. Polymers like PLA and poly(hydroxyl alkanoates) are biobased and biodegradable polymers that have the massive capability to significantly minimize the environmental pollution if used in packaging applications widely. Researchers are trying hard to eliminate the drawbacks of these polymers that hinder their widespread usage from commercial point of view. It is noteworthy to mention that some of the economically forward countries in the world have already started commercializing such biobased and biodegradable polymers like PLA to the market, specifically in food packaging applications. However, polymers like polybutylene adipate terephthalate and PCL are fossil-based biodegradable polymers that can also provide benefits to the environment. The high cost of production, insufficient thermal and mechanical properties and poor shelf life of most of the biodegradable plastics are the prime factors of hindrance to its multiple applications in the real life. It is expected that the mentioned limitations can be handled with the use of proper knowledge and understanding of the science and technology behind the material in near future.

References

[1] Rhim, JW., Park, HM., & Ha, CS. Bio-nanocomposites for food packaging applications. Prog Polym Sci 2013, 38, 1629–1652.

[2] Sorrentino, A., Gorrasi, G., & Vittoria, V. Potential perspectives of bio-nanocomposites for food packaging applications. Trends Food Sci Technol 2007, 18, 84–95.

[3] Valapa, RB., Pugazhenthi, G., & Katiyar, V. Effect of graphene content on the properties of poly (lactic acid) nanocomposites. RSC Adv 2015, 5, 28410–28423.

[4] Dhar, P., Tarafder, D., Kumar, A., & Katiyar, V. Effect of cellulose nanocrystal polymorphs on mechanical, barrier and thermal properties of poly (lactic acid) based bionanocomposites. RSC Adv 2015, 5, 60426–60440.

[5] Kathuria, A., Abiad, MG., & Auras, R. Toughening of poly (L-lactic acid) with Cu_3BTC_2 metal organic framework crystals. Polymer 2013, 2013(54), 6979–6986.

[6] Valapa, RB., Pugazhenthi, G., & Katiyar, V. Fabrication and characterization of sucrose palmitate reinforced poly(lactic acid) bionanocomposite films. J Appl Polym Sci 2015, 132, 41320.

[7] Tripathi, N., & Katiyar, V. PLA/functionalized-gum arabic based bionanocomposite films for high gas barrier applications. J Appl Polym Sci 2016, 133, 43451.

[8] Murariu, M., Dechief, AL., Bonnaud, L., Paint, Y., Gallos, A., Fontaine, G., Bourbigot, S., & Dubois, P. The production and properties of polylactide composites filled with expanded graphite. Polym Degrad Stab 2010, 95, 889–900.

[9] Schmid, M. Whey coatings for plastic films in food packaging. Adv Coat Surf Technol 2012, 25, 8–10.

[10] Oyama, HT., & Abe, S. Stereocomplex poly (lactic acid) alloys with superb heat resistance and toughness. Acs Sustain Chem Eng 2015, 3, 3245–3252.
[11] Dhar, P., Tarafder, D., Kumar, A., & Katiyar, V. Thermally recyclable polylactic acid/cellulose nanocrystal films through reactive extrusion process. Polymer 2016, 87, 268–282.
[12] Larsson, A., Seip, TAT., Storstrøm, O., & Fallet, T. High power module packaging design for harsh environments. iMAPS Device Packaging, Arizona, 2011, 2.
[13] Shia, Q., Zhou, C., Yue, Y., Guo, W., Wu, Y., & Wu, Q. Mechanical properties and in vitro degradation of electrospun bio-nanocomposite mats from PLA and cellulose nanocrystals. Carbohydr Polym 2012, 90, 301–308.
[14] Okubo, K., Fujii, T., & Yamamoto, Y. Development of bamboo-based polymer composites and their mechanical properties. Compos Part A 2004, 35, 377–383.
[15] Han, YS., Lee, SH., Choi, KH., & Park, I. Preparation and characterization of chitosan–Clay nanocomposites with antimicrobial activity. J Phys Chem Solids 2010, 71, 464–467.
[16] Yoon, OJ., Jung, CY., Sohn, IY., Jin Kim, H., Hong, B., Jhon, MS., & Lee, NE. Nanocomposite nanofibers of poly(D,L-lactic-co-glycolic acid) and graphene oxide nanosheets. Compos Part A 2011, 42, 1978–1984.
[17] Seligra, PG., Nuevo, F., Lamanna, M., & Famá, L. Covalent grafting of carbon nanotubes to PLA in order to improve compatibility. Compos Part B 2013, 46, 61–68.
[18] Xu, X., Yang, Q., Wang, Y., Yu, H., Chen, X., & Jing, X. Biodegradable electrospun poly(l-lactide) fibers containing antibacterial silver nanoparticles. Eur Polym J 2006, 42, 2081–2087.
[19] Pantani, R., Gorrasi, G., Vigliotta, G., Murariu, M., & Dubois, P. PLA-ZnO nanocomposite films: Water vapor barrier properties and specific end-use characteristics. Eur Polym J 2013, 49, 3471–3482.
[20] Chan, MA., & Krochta, JM. Grease and oxygen barrier properties of whey-protein-isolate coated paperboard. TAPPI J 2001, 84.
[21] Chan, MA., & Krochta, JM. Color and gloss of whey-protein coated paperboard. TAPPI J 2001, 58.

Index

https://doi.org/10.1515/9783110648034-013

Bei Fragen zur Produktsicherheit wenden Sie sich bitte an:
If you have any questions regarding product safety,
please contact:

Walter de Gruyter GmbH
Genthiner Straße 13
10785 Berlin
productsafety@degruyterbrill.com